OUR SERGEANT

OUR SERGEANT

The Story of the Corps of Commissionaires

by

PETER REESE

Leo Cooper
in association with
Secker & Warburg

First published in Great Britain 1986 by
Leo Cooper Ltd, in association with Secker & Warburg Ltd.
50 Poland Street, London W1V 3DF.

ISBN: 0–436–41003–6

Printed in Great Britain by
Redwood Burn Ltd, Trowbridge, Wilts

Contents

Acknowledgements

I acknowledge the gracious permission of Her Majesty The Queen for permission to quote from the Cambridge Papers and A. P. Watt Ltd, on behalf of the National Trust and Macmillan London Ltd, to quote from Rudyard Kipling's 'Shillin' a day'.

This book would not have been possible without much help from very many people. I acknowledge my prime debt to the present leadership of the Corps of Commissionaires. They kept faith with an unproven writer and one who wished to consider their Corps against that wider background of traditional struggle experienced by successive generations of servicemen re-entering civilian life.

Two organizations and their officials made fundamental contributions to my research. The Head Archivist of the Greater London Record Office, Miss J. Coburn, gave me ready access to the Corps' papers and Mrs Joan Kenealy, in particular, was immensely helpful. The Ministry of Defence, Whitehall Library, has helped me over many long years and in a wider context. I acknowledge the help of two Chief Librarians, Mr Derek King and Mr John Andrews, along with many others from this magnificent establishment, particularly Mr Charles Potts, whose encyclopaedic knowledge is staggering, and the late Stephanie Glover. Another library, the Prince Consort's in Aldershot, has ever given me ready refuge and assistance. Its current librarians, Paul Vickers and Tim Ward, maintain the tradition most ably.

Individual thanks are due to a number of others, two Commandants of the Corps, Colonels Geoffrey Pring and

Bruce Robertson for their courtesy and willingness in fielding my many questions and to their Personal Secretary, Mr Reg Codling. Mrs Frederica Sanderson gave me much valuable information about her distinguished family. There are those who helped bring the book into an acceptable form. Dr Leslie Wayper scrutinised the text with his characteristic incisiveness and thoroughness. He is surely responsible for many of the best ideas it might contain. Mrs Joan Gatfield typed the scripts swiftly and accurately from my almost indecipherable writing.

I am indebted to them all for their support and inspiration but any shortcomings and failings are, of course, mine alone. Lastly I thank my own personal assistant, my wife Barbara, whose sturdy common sense is infallible and on whose shoulders I have put so many contemporary burdens whilst scribbling and dreaming of the past.

CHAPTER ONE

The British and their ex-Soldiers: The Unacknowledged Problem of Resettlement

'But I'm old and I'm nervis
I'm cast from the Service,
And all I deserve is a shillin' a day.
Shillin' a day
Bloomin' good pay –
Lucky to touch it, a shillin' a day'

In those lines Kipling captured what has been the traditional British attitude to their retired servicemen. He was himself one of the forces inducing change in that attitude as his further lines

'Think what 'e's been
Think what 'e's seen
Think of his pension an' –
GAWD SAVE THE QUEEN!'[1]

so clearly show.

In times past, however, that had been an understandable attitude. It might be thought that a welshing bookmaker with the eyes of his clients upon him would be safer than those who bilk powerful armies of their pay. Yet it was often thought superfluous to pay them and not surprisingly soldiers turned to sack and pillage in lieu. It was the unpaid forces of the Emperor

Charles V that sacked Rome in 1527. Such practices, however, like everything else, had to be paid for and soldiers returning home found that their reputation made very difficult the reintegration into civilian life for which their habits in any case unfitted them.

Today the ex-soldier is a man with favoured qualities, who often possesses a range of skills and high self-reliance, one who is now fairly regarded and who generally succeeds in re-establishing himself satisfactorily in civilian life. Yet even now the phrase 'coming the old soldier' is used of one seeking to avoid hard work and responsibility and such current use is a reminder of the difficulties that had to be removed before the ex-soldier could be readjusted as adequately to civilian life as is the case today.

Being an island, Great Britain was able to fall behind most of her neighbours in instituting a standing army. She preferred to look to her moat for defence and was saying, even in the Tudor period, 'We care not for your soldier men who do the state disdain. But we care for your sailor boys who do the state maintain'. It continued to put its trust in its militia, the so-called 'trained bands' whose name suggested an expertise that nothing else about them would. Even these could function, however, only if there were men to train them, men who would spend most of their lives as soldiers. When they reverted to civilian life they had the problem of making a living and if they were no longer young and in good health this was an almost insuperable problem. This was recognized by Queen Mary Tudor who in her will directed that a convenient house be fitted up 'for the relefe and help of pore and old soldiers'. This was the first official attempt made in England to provide for the ex-serviceman and sadly nothing came of it. In the reign of the 1st Elizabeth there were economic difficulties as in the reign of Elizabeth the Second. Queen Elizabeth I disregarded her elder sister's wishes and left it to private enterprise to provide such facilities, her favourite, Leicester, making provision in 1571 for twelve old soldiers in an almshouse at Warwick.[2]

Under Queen Elizabeth, soldiers continued to be poorly regarded, both by herself and the population at large. In times of national crisis her semi-regular, trained bands were insufficient and extra men had to be raised quickly, often from the ranks of the unemployed, including rogues and vagabonds, although cases of actually emptying prisons were mercifully not common.[3] Due partly to such expedients the opinion of military life at the time was so low that employers were reluctant to re-employ returning soldiers, despite earlier records of loyalty and even if their places had not been filled. A handful of ex-soldiers received small pensions, but not the vast majority. Discharged soldiers were required by law to return home without delay where their home parishes were charged to assist them, particularly the married ones.[4] This served to make them far from welcome and considerable evasion occurred on the part of the parochial authorities.

The early Stuart period was hardly more favourable to the returning soldier and greater justice was not forthcoming until the period of the Commonwealth. Unfortunately, while this brought more help for such men, it also caused the people of the British Isles to harbour grave suspicions of a standing army and helped to perpetuate a degree of anti-militarism which was to cause continuing problems for successive generations of soldiers attempting to re-enter civilian life.

The seeds of antipathy and misunderstanding, which still obtain between certain Trade Unions and returning soldiers, were sown at this time by the granting of 'special' privileges to ex-soldiers. The main one was an Ordinance whereby men who had served for a period of not less than four years should be free to practise their trade or occupation in civilian life 'though they served not out their time with like immunities as if they had'.[5] There were no comparable concessions until the present century and then they were far less sweeping, being granted on evidence of the equivalent or usually more comprehensive trade training having been undertaken during military service. The Commonwealth Ordinance was one which could only dismay and anger the predecessors of today's Unions, the powerful Trade Guilds of the time, who saw the maintenance of training standards as one of their chief objectives.

Under the Commonwealth a traditional method of resettlement, practised by the Roman Army and others, was re-adopted: some soldiers serving in Ireland were allocated land from dispossessed owners. This caused a depth of bitterness which still plays its part in continuing troubles today. Admittedly it was a device used mainly to satisfy men whose pay was in arrears. But settlement tended to be a very protracted process and during the interval most soldiers, and even some officers, were forced to sell their land debentures at a fifth or two-fifths of their true value to rich officers or to land speculators.[6] Yet the granting of land to ex-soldiers was to be re-adopted as a method of resettlement in the following century, often with indifferent success. It was to be offered not just across the Irish Sea but in colonies where one could ensure the old soldiers were kept far from the mother country, largely in North America.

The Commonwealth also granted pensions for disabled soldiers and their families, a practice which aroused the envy of normal citizens who had no such protection against injury or disease at this time. This concession ceased at the Restoration.

Since Cromwell, like the Roman Emperors before him, needed to keep the loyalty of his Army, he was obliged to support his soldiers both during and after their service. He was not, however, concerned with resettlement as we know it today and, perpetuating as he did a reaction against the military, over the long term he made genuine resettlement more difficult to achieve.

Charles II, in whose reign we can see the beginning of the standing army in England, was as anxious as was Cromwell to do what he could for the soldier. He established a National Hospital for deprived old soldiers, in this following the example set by Louis XIV who founded the *Hotel Royal des Invalides* in Paris in 1670 and perhaps stimulated by the Marshal General of Ireland who planned to build a similar hospital at the Curragh.

The story of Chelsea Hospital is a colourful one, and it has been well recorded elsewhere.[7] The building of the Hospital owed as much to the King's Paymaster, Sir Stephen Fox, as to the King. He contributed no less than £13,000 of his own fortune to the project, almost three times as much as Charles's own contribution, most of which came from a fund allocated for secret service purposes.[8] To the King's eternal credit it should be remembered that Wren's splendid building was erected at a time of great financial stringency. Its purpose was to house 'such land soldiers as are, or shall be, old, lame, or infirm in ye service of the Crown'. The wording is significant. Like Kilmainham at the Curragh and the *Hotel Royal des Invalides*, Chelsea Hospital was not primarily for wounded soldiers, but for those who had served full careers in the King's Army and were then unable to obtain civilian posts.

Sadly Chelsea never succeeded in meeting its original aim. With successive increases in the British military establishment it became less likely that it could ever take all the old soldiers. The inmates, or in-pensioners, were soon to be outnumbered by an ever-growing number of out-pensioners provided with small pensions which in themselves were inadequate to live on. The out-pensioners were administered by Commissioners based at Chelsea Hospital and from 1692 onwards such pensions were funded by Parliament. This gave greater assurance against soldiers' pensions being discontinued at some future time, but it virtually guaranteed they would never be increased to a level high enough to live on with either dignity or comfort.

We cannot, however, see in the founding of Chelsea Hospital a genuine attempt to cope with the problem of resettlement. Indeed it was an attempt to deny that the problem existed. For it was believed at the time that soldiers should normally stay in the army for their whole adult life and that at the end of their service they could either be admitted to the Royal Hospital or given a limited pension. Soldiers released prematurely through ill-health or wounds were to receive pensions, although these were often temporary. The assumption throughout, however, was that soldiers would not be capable of re-establishing themselves in civilian life.

Something more worthy of the name of resettlement, although, like the mule, incapable of progeny, was the work of the impetuous, fiercely brave and warm-hearted John Manners, Marquess of Granby. His largesse to soldiers at the conclusion of the Seven Years' War resulted in his own impoverishment, and in his name becoming one of the most popular inn names in Britain.[9] He undoubtedly gave his soldiers a means of re-establishing themselves but the charity of a single individual could never be the basis of an adequate resettlement policy.

Nothing else, it seems, could do so either, at least until the nineteenth century. Yet long before then the necessity of some plan of resettlement would seem to have been clearly indicated. For Britain was at war for much of the eighteenth century. The army had to be expanded quickly when war came and reduced equally quickly when it ended. Men in greater numbers and of younger age then those retiring in peacetime were thus, at the end of the great wars, thrust back into civilian life to seek a living. They had to do so as best they could. Old practices were turned to. As in the days of the Roman Empire ex-servicemen were given the opportunity to settle overseas, but of new thinking there was little trace, and the assumption was still made that men from the army would not re-enter civilian life in the hope of competing successfully with those who had never joined the Colours. If the retiring regular did not enter the sheltering walls of Chelsea or Kilmainham Hospitals all he could hope to do was to find in menial labour a slight supplement to his small pension. And if this was the best that society could do for him the best it looked to him for was to make his continued existence as unobtrusive as possible.

He deserved better of it, for in emergencies his services were still required. He joined his still serving mates in the enforcement of law and order and, in the Plug Plot disturbances of 1842 and the Chartist troubles of 1843–4, the pensioner in his ill-fitting uniform and carrying his musket was frequently to be seen. And perhaps he was making things a little better for his successors than they had been for him. For before the Crimean War a greater proportion of ex-soldiers than had hitherto been the case seem to have found worth-while employment.

Even so the numbers of men leaving the service were still not large and the proportion of reasonable jobs obtained was smaller still. The great British public, it seemed obvious, continued to think poorly of the ex-soldier and of his capacity to contribute to civilian life. There is a Chinese saying: 'As the water runs the rocks are revealed', and as the years went by a pretty problem was unfolding. Until the ex-serviceman became more highly thought of the prospects of resettling him decently in civilian life would remain negligible. However, until it could be demonstrated that he could be decently resettled into civilian life it seemed entirely improbable that he would be more highly thought of, other than perhaps in the momentary exaltation of victory. Resettlement would be good for the self-respect of the ex-soldier. It seemed essential if he was to win the lasting respect of society.

The public's attitude to the serving soldier was changed by the Crimean War, a conflict which brought about a level of intimacy between the nation and its Army hitherto impossible before the invention of the electric telegraph. The evocative writing of W.H. Russell transmitted through the influential pages of *The Times* showed the inequality and harshness of British Army life as never before. Soldiers came to be seen in a new light, as extraordinarily uncomplaining, loyal servants of a country which gave them shameful support. 'Their heroic patience under suffering and privations have given them claims upon their country which will long be remembered and gratefully acknowledged.'[10] Through the pages of *The Times* and through her own tireless correspondence the country was to become acquainted with a heroine in Florence Nightingale, whose valiant acts were performed for soldiers suffering appalling agonies in the inadequate hospitals of the Bosphorous. The story of endless walks round the wards with her lamp, and of wounded men kissing her shadow as it fell did as much for the image of the soldier as for Florence Nightingale herself. Florence Nightingale's immensely strong personality and outstanding ability gave her access to men in the highest positions of state, like the Secretary of War, Sidney Herbert, and Lord Panmure, the Secretary of State, and the opportunity to influence them in reforming the conditions of serving men.

The new and uncharacteristic level of goodwill towards the soldier was carried to the middle classes from another source. In 1856 the Poet Laureate, Alfred Lord Tennyson, immortalized the disastrous unsupported advance of the Light Brigade in a poem which was read in hushed drawing-rooms the length and breadth of the country. This was a poem about a body of men led into a murderous trap, rather than a panegyric to their leader such as occurred in the Rev Charles Wolfe's poem *The Burial of Sir John Moore at Corunna* written during the earlier Peninsular War. While the deeds of British soldiers in the Crimea were still in the public's mind, British soldiers were to perform new heroics. Their swift and successful actions under General Havelock at the time of the Indian Mutiny were again material for W.H. Russell and *The Times*.

The Crimean War also played an important part in changing dramatically the attitude of the religious public towards the Army.[11] The evangelical *Memorials of Captain Hedley Vicars*, published in December, 1855, sold 70,000 copies in its first year. This demonstrated that a professional soldier could also be an ardent Christian. The stern and eccentric Baptist, General Henry Havelock, who died while conducting the relief of Lucknow, was made a national hero on his death and is commemorated by a statue beside Nelson in Trafalgar Square. He had not only been a good soldier, but was held to be a good soldier because he did not drink and because he loved Christ. With the rapid growth in Christian voluntary work among soldiers after the Crimean War, the term 'Christian Soldier' became common and the contrast with the picture of the hard-swearing pagan soldier of Napoleonic times very marked.[12] In a word, the Army became more respectable as a result of the Crimean War and the ex-soldier could be expected to rank also as a suitable subject for some form of charitable help.

That led to consideration of reform in many areas of military life. It highlighted the imbalance between officers and soldiers: it led to the abolition by Royal Warrant, since the opposition of the Lords prevented this being done by statute, of the purchase of commissions and the adoption of the continental practice where a competitive examination was the normal means of officer entry. It led to a consideration of the military shortcomings affecting the old Army and to the adoption of the continental system involving large reserves, which eventually produced an Army where most men served for between six and twelve years rather than the lifetime generally given to the old Army. Such a change was to lower significantly the age of the ex-soldier and to require him to be recruited and discharged in far greater numbers than hitherto. And this was to give a new urgency to the problem of reintegrating him into civilian life and to cause more concern with it to be aroused and greater effort to cope with it than had hitherto been seen.

CHAPTER TWO

Edward Walter and the Century of Reform

For centuries a society which thought little of its soldiers had demonstrated its lack of concern for them when they ceased to be so. An unexpected change in its attitude to serving soldiers was to give heart and opportunity to those who realized the importance of changing its attitude, too, to those who were leaving the Colours for civilian life.

It was, as we have seen, the Crimean War which brought about that change, for that was a conflict which brought about a new relationship between the nation and its Army, greater concern for individual soldiers and new efforts towards overdue reforms. For a long time that concern and those efforts did little more than make manifest how difficult the resettlement problem was to solve. For despite the new-found regard for the soldiers' exploits in war, the traditional prejudice against him re-entering civilian life was virtually as strong and persistent as it had ever been. Indeed it must be said that success was not truly forthcoming until the bulk of the male population had served during the two world wars and the second of these had been followed by the full employment of the 1950s and 60s. The persistence of the problem and the length of time taken to find an acceptable solution to it are a reminder of the great debt owed by present ex-servicemen, with their encouraging level of success in obtaining employment, to the reformers of an earlier age who often operated in a most discouraging climate of disinterest, if not of plain hostility.

One of these was a modest ex-gunner, W.G. Clifford, who did much to make soldiers more familiar with the true nature of the problems facing them on becoming civilians.[1] Dissatisfied with the shortcomings, even the hypocritical nature of official resettlement publications prior to the First World War, he wrote his own. This was a short, clearly-written guide entitled *How and Where the Ex-soldier Can Get Work*. Later he was to publish independently *The Soldiers' Employment Gazette*, a sixteen-page monthly guide which became a model for future military and civilian publications on the subject.

Another was Colonel (later General) E.F. Chapman, RA, who, as Military Secretary to the Commander in Chief, India (and later Quartermaster-General there), won over his chief to the case for an employment society to assist short-service soldiers. The result was the National Association for the Employment of Reserve and Discharged Soldiers, founded in 1885, by far the largest of the voluntary employment agencies and one which was to achieve semi-official status. Valuable as was W.G. Clifford's work in illuminating difficulties and suggesting the best means of dealing with them, it is obvious that his influence could not be as great as Chapman's.

Arthur Haggard, too, is worthy of mention. He was an author, like his more famous brother, Rider Haggard, but also a professional soldier who wrote on the lot of the private soldier. He was so moved by the difficulties and humiliations of old soldiers seeking jobs after the Boer War that he founded a Veterans' Club and was joint founder in 1908 of a major new employment body, the Veterans' Corps and Employment Bureau.

Four other prestigious reformers demand recognition. Within the punishing schedules of their professional lives these gave unstintingly of their energy to achieve a climate more conducive to better resettlement. One worked within the War Office; three were distinguished public figures.

Within the establishment Colonel Sir E.W.D. Ward, Secretary to the War Office and the Army Council at the time of Richard Haldane, still found opportunities to further his special concern with both the training and employment of soldiers and sailors due shortly to leave the colours. He chaired a number of committees of inquiry on the subject and was openly critical of the limited facilities offered at the time. In respect of training, between 1906–1914 he personally drafted guidelines for the provision and funding of technical training for soldiers, while in 1906 and 1914 the Committees he set up to consider employment-finding attempted nothing less than a co-ordination of the different and rival employment agencies for servicemen. Tragically the climate of opinion was not yet favourable for the recommendations of his 1906 Committee and the work of his later one was interrupted by the Great War. Despite his immense industry and unflagging determination, his ambitions remained largely unfulfilled in both areas, although he established a framework for training which others were to exploit.

The three public figures operated over the whole period following the Crimean to the aftermath of the Great War. The earliest, Lord Wantage, as Lt Loyd-Lindsay was awarded a VC in the Crimean War, and thereafter had a burning regard for the cause of the soldier, serving or retired, wounded or whole. He chaired the influential inquiry into the problems of Army recruiting in 1890 (which among other things identified the need for good resettlement), was a member of the Royal Patriotic Fund Commission, which dispensed monies to needy soldiers and their dependants, and was Chairman of the British Red Cross Society. His whole-hearted support for the National Association was just one feature, important as it was to resettlement, of this remarkable man's regard for the soldier of his age.[2]

Another steadfast champion of the serving and discharged man was Lord George Hamilton. As Chairman of the Royal Commission on the Poor Law and Unemployment (1905–9) he played a fundamental part in the foundation of national employment exchanges. Notwithstanding, he retained and declared his support for the ex-servicemen whom he considered, rightly, as being underprivileged in national terms. As long-time Chairman of the National Association, he guided it through its most difficult years immediately before and after the First World War.

The third great public figure who worked to change the climate of opinion towards ex-servicemen was no less a person than their Commander-in-Chief through three endless years of war. At the conclusion of the First World War Earl Haig took up no official post, as might have been expected, but devoted himself to the interests, particularly those relating to the employment and social welfare, of his less fortunate fellow soldiers and brother officers. His crowning achievement was the amalgamation of a number of voluntary bodies into the British Legion, one of whose functions was job-finding for officers.

All of these men, working throughout the period from the 1850s to the 1920s, helped to solve the problem of settling the ex-soldier satisfactorily in civilian life. But none gave more outstanding service in this respect than Edward Walter. As Wilfrid Owen was the poet of the First World War who above all others best caught the essence and the nature of the suffering then endured, so Walter of all those hammering away at the problem of resettlement most clearly saw and laid down the broad principles and guidelines which those who sought to tackle it must cling to.

At first sight one would not consider him a likely champion compared with the others. He served just ten years in the Army, retiring as a Captain. The Army List of the day showed him progressing by purchase from ensign to lieutenant in the 44th Regt of Foot (The East Essex Regt) and then again by purchase and exchange to Captain in the Eighth Hussars. He missed the chance to go on active service or to gain military acclaim, having retired in the year preceding the Crimean War. Yet Captain Walter was to exercise a seminal influence on resettlement which became both a basis and an example from which others could develop their own reforms.

While Edward Walter's military service was somewhat undistinguished he came from a remarkable family. His grandfather founded *The Times* and under his father as Chief

Proprietor it became the outstanding newspaper of its age. The paper was to gain even greater influence and fame under Edward's eldest brother, John Walter III, whose correspondent W.H. Russell had been responsible for the despatches from the Crimea which caused so much sympathy for the individual soldiers there, and which initiated widespread reform of the military machine and the conditions of service life. John Walter III was ever ready to use *The Times* in publicizing Edward's achievements in aiding the ex-soldier and sailor.

As boys the three Walters could hardly help being caught up in the affairs of *The Times* or affected by their father's relationship with it. As Chief Proprietor, it was virtually impossible for him to opt out of contemporary political developments, as it was the accepted custom for successive administrations to try to enlist the support of the newspaper. Whereas John Walter II had outstanding editors in Thomas Barnes and John Thaddeus Delane, as Proprietor, his own opinions could never be ignored by leading statesmen. His main interests lay in home politics and measures affecting the necessitous poor, some of whom were inevitably old soldiers, and contemporary Governments came to expect some comment, and most likely criticism, for their lack of scope and effectiveness in this area.

The Walters were a close-knit family and his sons must have been aware of the influence exerted by their father through *The Times*. An appreciation of the special responsibilities visited upon a family which owned the newspaper was second nature to them. They too must have sensed the endless currents of political ambition swirling around the topical lighthouse of the upper and middle classes. They surely shared some of the bitterness and anger their father experienced at his unexpected failure to enter the House of Commons in 1842 (John was 23, Henry 18 and Edward 17 at the time). Edward, for one, was to exhibit a hearty dislike of the party system, and the manoeuvrings which accompanied it, all his adult life. Like their father, the younger Walters had a high level of sensitivity towards personal affronts and a certain appetite for litigation. In the eldest and youngest there was to burn deep ambition in conjunction with the seriousness of purpose and religious conscience possessed by all three. All had a powerful sense of dynasty and strength of mind. Like their father they shared the need for periods of detachment within their family circles away from the public eye. He preferred to conduct the majority of the great newspaper's affairs from his country home rather than from Printing House Square. Again in the mould of their father, all three became implacably opposed to the pomp and outward show of the Christian experience adopted by the Church of Rome.

The careers of John Walter III and his youngest brother Edward have many clear parallels which do not appear immediately obvious. At first sight John Walter III's achievements were beyond comparison. Succeeding his father as Chief Proprietor on 1 July, 1847, four weeks after his death, he straightaway brought to his task abundant energy and strong convictions, combined with a sense of practical vision. He came to see himself as the heir of a great trust, a man with responsibility for the general political tone of the world's greatest paper, its English nature, even its faith and morals, an organ which represented the opinion of the country as a whole. To John *The Times* came to constitute a daily history of events, a national repository to which intelligent people would willingly go for the authentic day-to-day accounts. It had earned the right to investigate truth and to apply its own fixed principles to the affairs of the world, the principal ones being those of justice, humanity and law. During his long reign, lasting forty-seven years from 1847–94, John Walter succeeded in raising it to so high a standard that serious financial problems in the later years failed to weaken the paper's reputation.

The extent of John Walter's ambition and its practical implementation can be gauged through his main hobby – building. He designed and had built a country mansion which was very large indeed – an imposing red brick edifice four storeys high, called in jest, but not that unaccurately, 'the second palace of Berkshire'. His growing income from *The*

Times enabled him to build up an estate of no less than 7,500 acres, including a Public House appropriately named 'The Walter Arms'. The hobby became business when, after acquiring over several years a number of properties adjoining Printing House Square, he razed them and constructed a new and magnificent Public Office for both *The Times* itself and the Walter printing presses. This was opened in Queen Victoria Street in 1874. For this he acted as his own architect and all the bricks, 3,817,121 of them, were supplied from Walter's own kilns. The cost was borne wholly by him.

Even more than his father, John Walter III displayed a marked ability at management. He directed and managed the affairs of the world's foremost newspaper and in his scholarly way set his seal on its role and duty – always duty – to the Victorian state. He managed his employees, both on the paper and on his extensive estate, and he managed his own large family. He even found time to be a Member of Parliament for Nottingham from 1847–59, as his father was before him, and then for Berkshire from 1859–65 and 1868–85. By any standards his was a long and successful career and he had no false modesty about his achievements. In 1855, the leader, obviously with the proprietor's approval, described the paper as 'the most signal example of useful enterprise in the Empire'.[3]

One could forgive the youngest and less favoured son, Edward, for not competing directly with the heir to *The Times*. But Edward was to demonstrate that he too had inherited the salient and positive qualities of his father and could emulate those of his eldest brother. This was not immediately apparent in his early years, Walters were not generally extrovert, and not until shortly before his father's death had his eldest brother come to be seen as a potentially successful proprietor of *The Times*. Educated at Eton and Exeter College, Oxford, Edward made little academic or extramural mark on either institution. In both instances it seems likely his period of residence there was short. He attended Eton before there were registers to record accurately the pupils' stay. We find first mention of him coming low in school order in the summer list of 1835 and he must have left before class lists became a regular feature in 1838. He stayed at Oxford for a short time only, and, after matriculating in 1842, but not taking a degree, he went straight from there into the Army. Both his elder brothers had taken degrees at Exeter College. He appeared to enjoy his stay at University; like other undergraduates at that time, including his brothers before him, he ran up a number of debts, although his share was by no means as large as that of his second brother, Henry. Certainly John Walter II saw no serious flaws of character in the minor financial excesses of his sons and suffered them with wry humour. Indeed he considered them the norm and had neither much sympathy nor love for their Oxford creditors. Writing to his solicitor, Fred Soames, in 1843, he advised him to move slowly over payment 'The Oxford tradesmen are such rascals – Don't be punctilious with them.'[4] These same tradesmen were still submitting bills to careful John Walter in 1846, over three years after Edward had joined the Army. Carefulness in money matters was another trait of both father and sons in their mature years.

While his period in the Army was to have such great significance on his later life and work, little detail is known about Edward's military service save that it all took place at home stations, in England and Ireland, at a time when units generally spent much of their time overseas. Joining the 44th Regiment (the East Essex) in England he went with them to Ireland and exchanged into the more fashionable 8th Hussars just prior to the 44th's departure to Malta. Two months after selling out and retiring, he married Mary-Anne Eliza Athorpe. It was to prove a very happy union which lasted fifty-one years until his death. His wife came from a military and aristocratic family which had been founded by one of Cromwell's successful soldiers. One might be forgiven for thinking it owed much to early resettlement arrangements; the estate in Ireland originated from the purchase of soldiers' land grants. The military tradition persisted on Mary-Anne's side, two of her brothers serving in the British Army.

With one exception no documents survive from this period.

We do not know why Edward left the Army after ten years. Perhaps he had always envisaged a relatively short military career; perhaps he found, although undoubtedly a strong patriot, that certain aspects of military life were not to his liking. Perhaps he saw himself with limited prospects for promotion. Without firm evidence such speculation has little value. However, we do know he had the capacity to remain friends with a number of the young officers who continued to serve. They sent him warm letters from the Crimea. There is also a single document which helps to reveal another side to his personality at this time. This was a letter from Edward to his father in 1845, two years after he joined the Army. He referred to 'an affair at Devonport', a dispute between two young officers, Edward Walter himself, a Captain Browne and their Commanding Officer, Colonel Shelton.

With hindsight it now appears as the proverbial storm in a teacup. Captain Browne was apparently refused leave by Colonel Shelton on the grounds that Edward Walter, who had a prior claim, had insisted on taking his own leave at that particular time. Edward denied strenuously that he knowingly had any part in his friend being denied leave. He maintained he had been ordered to take leave at that time or lose the privilege altogether for that year.

Both young officers were plainly incensed and showed considerable courage and resolution in standing up to their Commanding Officer. Colonel Shelton had a long and arduous career, losing an arm in the Peninsular War and enduring much criticism for his conduct, but never his bravery, during the 44th's time in Kabul and its terrible retreat from Afghanistan. He was a peppery little man and even the Regimental History of the 44th admits that 'his temper and habit of mind were not his most endearing qualities'. A Commanding Officer of the time had sweeping powers, even over well-connected young officers, and this was the man they chose to oppose. In his letter to his father Edward said that Captain Browne had already written to the Commander-in-Chief about the mis-statement by his Commanding Officer and the gross insult it represented.

Hardly surprisingly the C-in-C was not inclined to rule against a Commanding Officer on such a minor matter and said he was not prepared to occupy the Public time by enquiring into the insult. Colonel Shelton appeared quite content to ride out what to him must have been a minor squall after his forty years' service. Neither of the young men was prepared to let the matter drop, though. Captain Browne considered retiring from the service if he could not get a Court of Inquiry to make the affair public, while Edward Walter wrote to his father strongly in support of his friend.[5] Edward saw the matter as one which degraded the basic military values of loyalty, truth and justice. There is little doubt that both young men took their soldiering, and themselves, extremely seriously. Edward's confidence was doubtless boosted by the knowledge that, with his father's help, he could invoke the influential support of *The Times*. Yet he was by no means bull-headed or without political awareness (this was not a Walter failing) and he respectfully promised his father he would 'take no step without, I hope, proper caution and the advice of those in whom I have every reason to depend'. In the event his father obviously considered it did not warrant further action and in early May Colonel Shelton was dead, the result of a riding accident. Yet the incident was instructive in demonstrating what a dedicated and idealistic soldier Edward Walter showed himself to be in 1845. Additionally it demonstrated he had the Walter flavour for possible litigation and a keen awareness of the powerful influence which his family could exert in the case of public injustice. It is significant that Edward Walter saw fit to preserve the letter, and it is revealing in that it showed an altogether more formidable side of someone who was referred to rather cavalierly in the following year by Crabb Robinson as 'A lively, agreeable lad'.[6]

In retrospect, Edward Walter's Christian military idealism, his devotion to the old regimental system, the service connections in both his own and his wife's family, the accounts of bravery and self-sacrifice concerning ordinary soldiers during the Crimean War related to him both in *The Times* and by friends still serving, his sense of regret tinged with possible conscience that he had not been there as well, all acted to arouse his sympathy and the Walter instinct for positive action in accordance with his own convictions. In fact, it could be said to have set the youngest Walter on his own crusade. Whereas his father had *The Times* and the necessitous poor and his eldest brother had *The Times*, which he raised to greater heights still, and his family mansion and estate at Bearwood, Edward was to champion a group of men newly recognized as valid objects of charity in Victorian eyes – soldiers, particularly those confronted with the age-old difficulty of rehabilitating themselves after their service. He was happily able to pursue his own practical vision for almost as long as his eldest brother was to guide *The Times*. In his own way Edward was as much a despot in his affairs as his brother and he also practised the necessary detachment all Walters required within the bounds of their own families.

The Walter single-mindedness was to give him success which was not inconsiderable compared with the other two. Edward managed his men with an unsentimental but true touch, reminiscent of, but no less successful, than his father's and brother's relationships with their employees. His highly shrewd programme of property acquisition for his old soldiers reflected his eldest brother's profitable regard for bricks and mortar. And whereas *The Times* represented a pinnacle of journalism, a peak never surpassed by the mass newspapers which followed, Edward's body of old soldiers set a standard of conduct and a reputation for trust never before attributed to the archetypal old soldier and never again equalled by other military employment associations. Moreover in one important respect he outstripped his older brothers. His work brought him a greater public acclaim than did theirs and his knighthood was a signal honour never before received by a Walter.[7]

CHAPTER THREE
The Mission Unfolds

As a retired junior officer, Edward Walter could expect to exert little or no influence on the serving Army of his day. Yet as an individual with powerful connections and some financial standing, highly committed to a long neglected cause, he could hope to play a premier role in the resettlement of the ex-soldier into civilian life. It was hardly surprising, then, that he should be involved with, and was most probably the inspiration for, the significant initiatives taken on behalf of the old soldier, both during and after the Crimean War. One might almost say it was an inevitable pattern emanating from the Walter sense of ambition combined with a dedicated and powerful sense of moral justice. Any such remark, though, would give hardly less than a just acknowledgement of the positive steps taken by a young man who had as yet to demonstrate his particular force of personality among his older and immeasurably senior colleagues.

He was a founder member of the first organization ever formed to help old soldiers and sailors find employment. This was the Army and Navy Pensioners' Employment Society established specifically to help the large numbers of servicemen discharged during the Crimean War. Such men were unfit for military duty both on account of their wounds and because of broken health due to the rigours of the campaign. Almost without exception such men experienced the greatest difficulty, almost amounting to impossibility, in helping themselves and their families. They were not in receipt of adequate pensions and therefore needed work with a reasonable level of pay. In consequence of the problems they faced, many of those men lost heart and became half-destitute; even worse in the eyes of their ex-officers (including Edward Walter) they rapidly lost the core of their existence – their military instincts and habits of discipline and order.[1]

A number of 'gentlemen' with consciences, the majority of them having military connections, met at the London Tavern on 18 July, 1855. Their purpose was to establish an employment society for all men of good character, from the Army and Navy, who qualified for a pension. They were instrumental in sending out a circular to prominent public figures in order to publicize and enlist support for its work. This offered graduated membership: for five guineas a person was given life membership (and the privilege of recommending men for situations), for one guinea, annual membership, with the same privilege regarding recommendations. All regiments or ships interested in the chance of recommending men for situations were required to pay £2 annually. Her Majesty the Queen was approached and agreed to support the Society through an annual subscription of £50, while the Prince Consort matched her contribution. Support also came from the Duke of Cambridge, the Commander-in-Chief, from 'many noblemen and gentlemen', even, it seemed, from 'some benevolent ladies'.

The problem, however, was both more long-standing and more intractable than most of these benefactors realized. It had been highlighted by the exceptional numbers arising from the war and it would remain long after that war had been forgotten. Indeed the number of applications for employment would remain high, for soon the Indian Mutiny caused a rapid augmentation and subsequent rundown of both the Army and Navy. While one can now see strong reasons for making the Pensioners' Employment Society a permanent institution, at the time the feeling was far from general that this should be done. Indeed the enthusiasm of the many leading figures who came together during the Crimean War soon cooled at its end.

This resulted in a serious fall of income and by 1859 the society had fallen into considerable debt.

While Edward Walter had the most modest military qualifications among those who came together in 1855, he was to prove the key figure in the Society's survival. It would surely have been dissolved if he had not held himself personally responsible for its liabilities. The young ex-Captain was now to show the qualities of unyielding determination, cool confidence, organizing ability and a mastery of detail which were to prove so important in his later contributions to resettlement.

The main condition he stipulated for rescuing the Society was that henceforth the Executive Committee would relinquish its powers to a Council whose membership was limited to Officers of Her Majesty's Services. The military were henceforth to look after their own. The Council was to be assisted by the one full-time member of staff, a paid Secretary. The revised rules of 1860 clearly owed much to Edward Walter; they followed the general pattern of his regulations for other bodies, including the Corps of Commissionaires.

The Secretary's duties were extremely detailed and they were specifically laid down. He appeared to have very little chance of allowing the organization to fall into debt again, every item above the amount of 2/6d (12½p) having to be separately specified in the Annual Statement of Accounts. Membership of the Council, a quarter of whom retired annually (although they were eligible for re-election), was no sinecure either. Members were required to take weekly turns at superintending the general business and detailed working of the Society – the roster book being kept by the Secretary. The senior officers concerned must have thought that to some degree they were re-enacting their days as young orderly officers. At the very least they must have been in no doubt the system had been devised by an enthusiastic junior officer. Some of the wry remarks which such a situation invoked can well be imagined. If Basil Liddell Hart was the Captain between the World Wars who taught Generals, Edward Walter was certainly the ex-Captain after the Crimean War who tasked ex-Generals.

One has no difficulty in detecting, through the declared objectives of the revitalized society, the strong moral commitment of Edward Walter towards the cause of the old soldier. It was now dedicated to perpetuate and even to enlarge its sphere of activity. Whether in peace or war the Society was to act as a means of raising the character of the Services by demonstrating to the soldier or sailor that 'in the decadence of his powers, whether from wounds, climate, accident, or long service he was still cared for'. With proof of good character any old soldier or sailor was eligible to claim its aid, although those recommended by contributing members would enjoy some advantage.

The achievements of the Society were considerable (between 1859 and 1873 nearly 9500 eligible Pensioners had registered and 7573 been provided with employment).[2] Yet it had distinct limitations. The great majority of employers' applications were for servants and, although Edward Walter and the Council wanted particularly to help those men who had no trades, they met with disappointing results: many of the employment situations dealt with were temporary ones and quite poorly paid. Also discouraging was the fact that, despite every effort being made to attract increased financial support through encouraging donations from individual officers, their Regiments and Battalions, funds remained very meagre.

In the same year that Edward Walter saved the Pensioners' Society he went further and took the momentous step of establishing his own organization to help old soldiers. For, despite his premier role in it, the Pensioners' Society, as constituted, could never satisfy him. True, he could organize the Council's participation in its day-to-day workings; he could also exhort its members to prevail on their military colleagues to help financially. Yet in his heart he must have known that this could never become the vehicle of his grand design which was nothing less than breaking down the deep-seated prejudice against the old soldier. To succeed in doing this he believed that he needed to do two things. He had to persuade Parliament to give ex-servicemen vacancies in public departments, and he had to provide dramatic and incontrovertible proof that such

Captain Edward Walter (1824–1904), Founder of The Corps of Commissionaires.

The Original Eight.

An early commissionaire. His medals indicate that he once probably served in the Bengal Sappers and Miners.

Commissionaire Mara, late 41st Fo
(The Welsh Regiment).

men could perform a valuable service to employers.[3] He believed that ex-servicemen would be of most help to employers through the qualities and virtues developed during military life rather than through the deliberate acquisition of civilian-related skills. He knew full well that the majority of cavalry and infantry soldiers had little opportunity to acquire such skills. The Pensioners' Society, with its modest employment posts and relatively small numbers, could never accomplish this double task. Its men were spread haphazardly in single numbers, as indoor and general servants, as porters and watchmen in warehouses, as railway carriage cleaners, as 'boots' in hotels or grooms in their livery stables. Through these roles many lost their individuality as old soldiers and sailors. If they did succeed in keeping their identity as archetypal military men, they often became embodiments of the scant regard in which the ex-serviceman of good conduct was held. What they were doing would never convince Parliament of the desirability of securing for ex-servicemen places in the Civil Service.

Walter's own creation set out to change this state of affairs. In the first instance he tried to resettle the most difficult cases of all, those who had been wounded. The eight original members of the group he now founded, seven soldiers and one sailor, had each lost a limb, but through friends and contacts he soon found employment for them all. As evidence of his deep Christian convictions and faith in God's support, on 13 February 1859, he took them to Westminster Abbey and gave thanks for the early success of his movement, his future life's campaign. To Walter it was not enough that his men had work. The nature and means of their employment were equally important. He looked at the assets possessed by an ex-soldier of good character, his military bearing, the ability to set off a good uniform and earn public respect, his integrity and trustworthiness in carrying out a commission. Walter knew full well that ex-soldiers and ex-sailors could guard a civilian strongpoint, be it bank or office block, with patience and effectiveness, using no weapons save the authority of a suitable uniform and their own purposefulness and courtesy. Walter's genius lay in his understanding that it was in their roles of guarding and message-carrying that 'ordinary' soldiers could best serve the City of London, or any other great Metropolis, and transform the opinion held by the public of ex-soldiers in general. Unlike the men of the Pensioners' Society, Walter's own men would not be hidden away. They would be on show, noteworthy through their distinctive uniforms striding along public streets and occupying the portals of the great corporations. Such men would not be seen relying on charity of any description; they must demonstrate Walter's belief in self-help and thrift. Ex-servicemen should not be objects of sympathy and compassion like the common poor.[4] He knew, in justice, that what ex-servicemen needed most were special employment opportunities, after years away from the civilian community, and he attempted to obtain posts within the Civil Service and the great Departments of State. In the meantime and to further this end they needed good employment in order to present a more favourable image to the Public and to successive Governments. He believed they could demonstrate through their military qualities the essence of their employability to the City of London, the world's financial and commercial capital, to the country as a whole and indeed to the Empire beyond. The City would surely come rapidly to appreciate representatives on the doors of Head Offices who embodied discipline and loyalty, men who could guard banks and storehouses, men of trust who could carry sensitive and valuable items between branches of companies and between companies themselves. In short Walter needed ex-servicemen who, while not possessing specialist skills, were of superior stamp, of unimpeachable honesty and who in their uniforms were always exposed to the public gaze. Thus exposed, they had to be able to impress.

Not surprisingly Walter's organization was to experience many struggles and setbacks in pursuit of his idealistic goals. Nevertheless, without success here he could not have hoped to persuade leading figures to support his complementary and greater aim, the securing of places reserved for ex-servicemen in the public Departments of State.

CHAPTER FOUR

A Struggle for Survival, 1859–1864

By the end of the 1850s Edward Walter had decided what required to be done for ex-servicemen. Primarily they needed the chance to take up work opportunities in the civil departments as of right, departments like the Customs and the Post Office. This was the long-established system on the Continent, particularly in Prussia and France.

Walter had also determined how he might put pressure on successive governments into granting such concessions. He planned to work through influential people who were capable of furthering the cause and to keep alive a running debate both in Parliament and other significant forums, like the Royal United Services Institution. He knew he could always rely on sympathetic and regular support from *The Times* in helping to keep the subject before the public eye, and later he personally published a compendium of debates and articles upon the subject. In particular, he knew that if he could form an organization, and through it could prove the serviceman's sense of responsibility in the field of civilian employment, his campaign would have markedly greater momentum and effect. The activities of the organization could, if successful, give him the greater status and significant opportunities needed to pursue his aims from a level of responsibility and involvement never achieved during his early life.

The Continent not only offered him the precedent for granting ex-servicemen reserved posts in government, France was to give him the idea and title for his pioneering organization. Attached to every hotel in every large town and procurable at a few minutes notice were *valets de place* and *commissionaires*, the former of whom would conduct the stranger to all objects of interest in the neighbourhood, while the services of the latter were useful in getting the baggage safely through the customs, carrying letters and running errands. He therefore called it the Corps of Commissionaires, anglicizing the word *commissionnaire* by taking out the second N.[2] It was an inspired choice, the foreign title had a ring about it, a dignity and air of mystery removing it from the normal connotations given to messengers, doormen or attendants at that time. It had a cachet far more impressive than the humble yet long-winded title accorded to the Pensioners' Employment Society. The word 'Corps' again had greater significance than 'Society'; it denoted something of a military nature, an appropriate symbol for the disciplined and unique body he had decided to form.

Strictly speaking, Edward Walter had already taken initiatives in employment-finding before giving his embryo organization a title. He had advertised for the services of wounded men in need of jobs and placed all eight of those who replied. Sadly, the ability of severely wounded men to perform the more demanding tasks was to prove somewhat limited, although the Founder never ceased to declare a preference for such men to join his Corps.[3] The 'original eight' must also have fallen short of his ideals as far as dress was concerned. From the group photograph of the 'originals' (see Fig. 2) one might easily be forgiven for confusing them at first glance with those ex-soldiers from the US Civil War who feature in the archetypal Western. This image was far from that of the proud commissionaire conceived by Edward Walter. To help make the latter a reality he created a uniform which would exemplify the dignity and officialdom of their role. It was to be dark green and equal in quality with that of a sergeant in the Rifle Brigade. There were to be winter and summer trousers, each with red beading down the legs, a forage cap, water-proof cape and greatcoat and, most important, a cross belt with pouch, the mark of the commissionaire.

To further his campaign for reserved posts Walter needed to make his Corps succeed; indeed, it had to attain the high standards he set both by reason of his personal ambition and to enable him to achieve his overall objective. Three contradictory factors ensured that the Corps' early days would be ones of great struggle, at times bordering on despair; these were the Founder's idealistic and unyielding code of conduct, the fallibility of contemporary ex-servicemen and the limited resources available.

The commissionaires' required code of conduct was embodied in the Rules and Regulations which he drew up and produced in full printed form in March, 1864. More will be said of these later, although they could best be described as an amalgam of the Corps' Army Act and its Bible, representing the way to security and salvation for individual commissionaires and for servicemen's resettlement as a whole – a text which must be followed to the letter.

Of paramount concern in those early days was what shape the Corps would take. During 1859 the Founder obtained rented accommodation in Central London, in Exchange Court just off the Strand. The building was in the main of timber construction and quite modest in proportions but its position was most suitable when one considers that the City of London was the main employment target. He called it The Barracks, demonstrating his intention of forming no ordinary employment organization but one in the form of his own private army. Unlike regular army barracks, and his own organization in later years, it was as yet unable to offer living accommodation to many commissionaires. This affected the degree of control over individual commissionaires and the extent of his flexibility in despatching men on commissions at the shortest notice and at all hours.

Given the primitive nature of the barracks it was only to be expected that the Founder did not himself occupy such office accommodation when administering the Corps' daily routine. In any case he would inevitably have to be called away from time to time to deal with wider issues. Edward Walter therefore based himself at the Army and Navy Club and delegated the execution of local matters to a man appointed as Company Sergeant-Major who was to be assisted in the office by one other.

From the Sergeant-Major's letters one realizes the disadvantages of this remote control. The Founder gave him the most detailed written instructions, because he was in largely uncharted waters as to the direction his organization should go, because he did not delegate very easily and because the Sergeant-Major was not himself up to taking on much individual responsibility. An extreme example was the Sergeant-Major's inability to spend 2/6d on a London Guide, a necessity for new commissionaires, without express permission to do so by the Founder.[4] The first Sergeant-Major showed all too many of the old soldier's qualities. He could not resist emphasizing how hard he worked, 'On Thursday last I walked over about thirty miles of ground', and was ceaseless in his complaints about his fellow helper in the office, a commissionaire named Murphy. From successive reports it appeared that Murphy was lazy, unreliable with the accounts and prone to drink.[5] He was subsequently fined and dismissed, the main surprise being that he survived the Founder's wrath for so long. From their correspondence it was also apparent that Edward Walter did not always accept his Sergeant-Major's comments unequivocally.

The devolved system of control continued through 1861, although two officials now acted in a position senior to that of the Sergeant-Major. These appeared to be ex-officers who were given the title of Secretary and by the end of that year the military post of Adjutant had certainly been adopted. The communications from Headquarters to the Founder during 1861 were less subservient, despite the Founder's instructions being as detailed as before. Yet they clearly illustrated the serious range of problems being experienced. Many concerned the calibre of the individual men. Apart from the ubiquitous Murphy who continued to feature regularly and who was clearly acknowledged by one Secretary to have sorely tried the Sergeant-Major, others were prosecuted by the Founder for dishonesty and it was considered that several were plainly not good enough on many counts. Dismissals were frequent but it was thought that better men were beginning to come into the Corps. One letter to Edward Walter gave an instance of an extra post becoming available due to the success of a reliable commissionaire – hardly conclusive evidence yet. On the other hand wounded men were proving themselves incapable of performing the more demanding duties of the time; not surprisingly, one-legged men were not up to standing for any length of time and for portering duties one needed a man with two arms.

It was evident from these early reactions that men with special qualities were required; many of the servicemen discharged with very good character references were found in fact to have grave weaknesses when given tasks involving a degree of individual responsibility. This could only be of the gravest concern to the Founder who was in the process of creating a body based on those very qualities of responsibility, unimpeachable honesty and service. It could only result in a large rejection rate of old soldiers, some of whom raised public criticism by their vociferous cries against unjust dismissal, and, by reason of these dismissals, limited the growth rate of the organization. It explains to some extent why Edward Walter's rules and regulations appeared to be moulded in iron, allowing for no latitude or personal discretion. He was adamant in maintaining high standards, for would-be commissionaires had to accept that they were operating at all times under the critical gaze of the public, although he came to offer 'these men apart' special benefits too. Essentially he provided an environment similar to their service life, operating a system of special controls and privileges but offering them 'a regiment to go back to' in which they could feel at home. Firm emphasis was laid, as in all military units, upon the common interest, and a man's misconduct was not only wrong per se but, most serious of all, because by it 'the character and credit of the whole Corps will be compromised'. Fundamental to the new organization was the way in which ideas could become reality as far as job opportunities were concerned. In this respect a certain pattern of employment was discernible as early as mid-1860.[6]

Edward Walter's commissionairing system offered members of the public the opportunity of employing an individual or group of men to execute certain duties on their behalf, notably those of guarding property or carrying out various commissions. At best Commissionaires were permanently employed, but as yet these were few in number; some twelve served under Government in the Department of the Royal Engineers and Commissariat while there were others in private establishments. The private establishments included domestic households, a notable example here being the home of Florence Nightingale, but they also included clubs, hotels and shops.[7] The men in private households were seen by relatively few, if influential, people, although those in clubs and shops could publicize the Corps' image most effectively. The same could be said for the few commissionaires employed permanently at main railway stations, including Belfast, Cork and Edinburgh. Here they could not avoid drawing the attention and, hopefully, the gratitude of the many railway travellers. If not on permanent employment, and most were not, commissionaires stood available for duties in two main locations. They were ready to be sent from the barracks on assignments to any part of the United Kingdom, usually for the period of a day (cost 2/6)* and from their depots and posts along the London streets for shorter periods still.

At this stage the front line of Walter's Army constituted the manned posts on the streets of London, supported as they were from their district depots and ultimately from the Headquarters itself, The Barracks. Here the majority of commissionaires stood in all weathers awaiting tasks of fetching and carrying. They publicized the Corps and if they performed their temporary tasks well it was hoped that more permanent places would follow, both in Government and private employ. Employers would pay according to a set tariff: if by distance it was 2d for half a mile, 3d from half to one mile. If payment was to be made by time, the rate was 2d for a

* To avoid the necessity of giving the equivalent figure in decimal currency on each occasion, younger readers are reminded that there were 20 shillings and 240 pence in the old pound.

quarter of an hour, 6d per hour, working on the assumption that Commissionaires walked at 2½ miles per hour. For every parcel over a stone there was a premium of 1d a mile. On taking up an assignment and receiving the tariff the Commissionaire was required to submit a ticket to his hirer on which the price of the tariff was recorded. When given a commission each man had to report the fact to his principal post, or depot, and, if possible, to his nearest comrade, stating the probable time of absence. No man could take any special engagement without permission from his depot. This system ensured the men's continuous presence and availability to the public. It was an interlocking and orderly system. Commissionaires without tasks were required to leave their posts for a short time and assist their depots in case messages had been sent down there from head office. When men received a direct commission their depot gave cover in the event of an absence of any substantial duration. During strictly temporary absences arrangements were normally made for shops in the post's immediate vicinity to accept messages and parcels. These had to be addressed to the Secretary or Sergeant-Major of the Corps. The system was also effective in the distribution of circulars. To any particular district the rate for delivery was 4/- per 100, delivery from house to house cost 1/- per 100.

The individual commissionaires on post duty worked to a highly controlled régime. Prior to taking up their posts they attended morning parade in the Barrack yard at 0830 in summer and 0900 in the winter; exceptionally they reported earlier still if their assignment was before parade hours. They were required to take up their posts immediately after parade and any messages given had to be carried out with strict regard to the priority of order. Apart from periodic visits to depots for any tasks allotted by headquarters they could not leave their posts until the end of their shift. Men had to carry out their tasks as quickly as possible and not perform any unofficial ones. On inquiring the way a man had to apply to a policeman or a respectable shop but on no account was he to enter a Public House and so greatly did Edward Walter fear the dangers of strong drink on his body of veterans that if anyone did so he was subject to dismissal. (It was hoped that commissionaires would abstain at all times from frequenting public houses.) Any case of drunkenness was punished by instant dismissal. No commissionaire could trespass on the post of a comrade but in that man's temporary absence he could undertake any commission offered by the public. The regulations did not only constrain, incentives were there too; the well-conducted men were given all the vacancies in desirable posts and all appointments to permanent stations were given to commissionaires of the best character and smartest appearance. Besides, a distinguishing badge was devised so that the Public would recognize and give preference to such men.

There is no doubt that thoroughgoing rules were necessary in situations where men were open to considerable temptation. They handled considerable amounts of cash when fees were paid directly to them (although they were required to submit tickets for the amounts paid). The first Sergeant-Major experienced the strong reluctance of men to move from their own pitches, for after working these up many came to receive regular perks from their habitual users and even some from clients who wanted preferential treatment in the carrying out of their commissions.[8] In July, 1860, the Founder had published a single newsheet to publicize the essential facts about his Corps; this set a confident tone regarding the growing size of the organization since its establishment eighteen months before. From the opening sentences one could be forgiven for thinking that it was just a matter of time before commissionaires took up a range of Government employments, penetrated most of the important private establishments in the country and provided a comprehensive service countrywide.

Such confidence was essential in a broadsheet of this nature. For obvious reasons it did not emphasize the struggles for survival which had occurred over the same period. The nature of these could still be gauged to some extent by a closer examination of the broadsheet information, particularly its opening sentences.

A heroic note was struck from the outset declaring that 'the Corps was composed of Soldiers and Sailors who have served their country in various parts of the world, and, from the wounds they have received, are precluded from ordinary occupation'. By this time, though, it was likely that the preponderance of men had not in fact been maimed by war, although there was no doubt that service overseas undoubtedly took its toll of many constitutions when the men came to pensionable age. So too did their personal predilections which for many had involved hard and prolonged drinking. The following sentences re-emphasized the commissionaires' personal qualities, their independence and, shrewdly, the moderate tariffs they demanded.

If in the cause of attracting continuing support the Founder emphasized the wounds his commissionaires had sustained more than their relative debility caused by harsh climates and their way of life, his emphasis concerning their reliance on self-help and their own exertions minimized, indeed ignored, their special financial sacrifices to this end. It ignored the all-embracing nature of his disciplinary code which ensured that their conduct was indeed exemplary by his own stern standards. No man was able to retain all his earnings. Along with his fellows he suffered considerable stoppages in order both to support the body of the Corps and to perform adequately his role as a commissionaire. On joining, all men were required to make an initial deposit of £1 as a guarantee of good conduct (to be forfeited in the case of dismissal) and a contribution of 5/- to the Corps General Fund. When serving, they paid 1/- each month to the General Fund. For every Permanent Employment received they made a contribution equal to one week's wages and emoluments, for every temporary employment received from Headquarters a contribution of 1d per shilling out of net wages.

The quality and specifications of the uniform and equipments worn by commissionaires were determined by the Commandant. No article of clothing or equipment was the private property of any individual commissionaire who was strictly forbidden to lend or dispose of any part of it. In the event of his leaving the Corps he was required to return the whole into the Quartermaster's store (to be sold and credited to the Clothing Fund). All men needed to have a complete set of clothing at all times and articles deficient had to be paid for at their probable value when lost. Complicated rules operated for losses and returns but each serving commissionaire had to pay 1/6d weekly into the Clothing Fund for his uniform in addition to his other liabilities.

Two other demands were made on a commissionaire's wages, the first for health and welfare reasons. Every man joining in a good state of health, who was not a member of a certified Sick or Benefit Club, had to pay 5/- entrance money and 3d a week, payable on pension day; men of impaired constitution were to pay 1d a week extra. This entitled every man at best, in the event of sickness, to receive 7/- per week for four months and half this amount for a further two months. At age 65 a commissionaire ceased to belong to the Sick Fund but up to £2 10/- would be available from it towards his funeral expenses.

The second liability concerned his personal and family welfare. The commissionaire was required to put not less than 1/- a week into the Corps Saving Bank against the time he might be unable to continue working. He was not allowed to withdraw any portion of it, save the interest, without special permission. In the event of his resignation or dismissal it was repaid within a fortnight.

Added to these compulsory stoppages was the substantial risk run by commissionaires in suffering fines. These were stringently imposed to keep that discipline which to the Founder was the very essence of his Corps. The most serious financial punishment – a fine of 2/6d – was for two possible offences, absence without leave when detailed to attend the funeral of a comrade, or for giving an incorrect address if allowed to live out of barracks. Absence from Muster Parade warranted a 2/- fine and from Church Parade one of 1/-. General Church Parade took place on the second Sunday of each month from which there was no exception (on other Sundays married men of more than a year's service could obtain permission to accompany their wives to a church of their choice, providing their Minister forwarded the application and on the understanding he would report any instance of non-attendance). Commissionaires had to be practising Christians and seen to be so by the general public. Even if a commissionaire had been given permission to wear civilian clothes by his employers, he was required to wear uniform on Sundays 'in order that by the publicity thus afforded, increased opportunities may arise for the employment of his comrades'.[9] He also had to be smart at all times, was not to lounge about his post, wear his hair according to the custom of H.M. Services and to have it cut at least once a month. To smoke while out on the streets in uniform detracted from such military standards and was thus liable to a fine of 1/-. Other misdemeanours were less expensive but they still reflected the high level of control imposed; being late on parade 1d, more than five minutes late 2d, being untidy in appearance 3d (and reported to the Commanding Officer), neglecting to pay the debt due to the Corps for clothing etc on pension day 6d, on the following day suspension.

Apart from the initial payments on joining, commissionaires were required, therefore, to pay no less than 3/- a week into the various funds specified by the Founder. Unquestionably there were good reasons for the level of stoppages imposed. Any lower rate of subscription into the Sickness Insurance and Savings Funds would have made their ultimate benefits very questionable. The uniforms selected by the Founder were of good quality for both practical and prestige reasons (they were bought in bulk to reduce the cost), and the financial soundness of the Corps certainly required no less a subscription than the weekly 1/6d per man. Each commissionaire's contribution of 3d a week towards the General Fund, which provided the Corps' running expenses, was little enough by any standard. Yet by 1864 a commissionaire's average wage was 21/- a week and it had been rather less three years earlier. Like all averages some exceeded it and others were below. The result was that men were paying at least 14% of their weekly wages in compulsory stoppages. This calculation does not include the two other considerable payments already mentioned; the contribution equal to one week's wages due from men found permanent employment (obtained for them through Headquarters), or a contribution of 1d in every shilling earned by those in temporary employment (again obtained from Headquarters). These temporary employees were likely to earn below the average and therefore paid something in the order of 25% from their weekly wages.

The Founder made every effort to reduce the degree of hardship accruing from such stoppages by requiring payments on Pension Day (Quarterly) when they could be deducted from this source rather than from weekly earnings. Nonetheless such stoppages laid the Corps open to scurrilous and potentially damaging attacks from certain popular newspapers of the time such as the *Star* and the *Comet*[10] and even to criticism from influential men like Charles Gordon (Gordon of Khartoum) who was still remarking unfavourably on the commissionaires' level of wages some few years later.[11]

In criticizing the Founder so vigorously, both for the strictness of his code and the stern and inhuman way he applied it, the newspapers had little idea, and less interest, in finding out how fallible these early commissionaires were. The Corps gave them a Godsent opportunity to snipe at *The Times* and the high moral and charitable stance it so evidently adopted.

The extent of the commissionaires' fallibility can be discerned from another self-congratulatory statement in the 1861 Broadsheet. This was to the effect that nearly 200 men joined that year, but after five years' existence the Corps still only numbered 250. This was not primarily due to a shortage of volunteers but because the wastage through resignations and dismissals over this early period was very high. While insisting on the highest standards, Edward Walter, of all people, could hardly have been content with the relative stagnation of his young Corps in manpower terms.

Another fundamental problem facing the Founder at this time, which was given only the briefest of mentions in the

Broadsheet, along with the need for commissionaires to help themselves, was the entirely self-supporting nature of the institution itself. Edward Walter had already experienced and sorted out the financial problems of the Veterans' Society. He knew well the penalties for lack of vigilance, imperfect financial control and waning public support. When embarking on his own more ambitious project he attempted to ensure that, for the most part, it would rely on its own resources, on capital which was self-generating. This itself imposed severe growth limitations during the initial phase and made his control appear harsher than it need have been. He just could not afford to subsidise individuals from the wafer-thin resources of the General Fund. In certain instances he even contacted the Paymaster of Pensions requesting that commissionaires' debts should be paid from their future pensions. The reasons were twofold: men must not be allowed to get away with protracted debts and the Corps needed the payments. As the 1864 Rules explained: fines were not merely to punish a man for committing an offence against the general welfare and interest of his comrades, 'but to compensate the Corps in some degree for the trouble and expense entailed by the necessity of keeping an extra Staff for the maintenance of good order and discipline'.[12]

The financial stringency also explains how certain of the rules regarding monetary matters might be interpreted as being weighted against individual commissionaires who were without redress because the regulations were inviolate. The sternest critics, however, could not deny that the Commanding Officer was as punctilious in his conduct of the General Fund as in all other matters concerning the Corps. He appointed five commissionaires to administer the Fund, two chosen by himself, two from the general body of commissionaires, the last being the Senior Non-Commissioned Officer. They were obliged to meet every week without fail.[13]

The General Fund was liable for the payment of all expenses incidental to the working of the commissionaire's system. Edward Walter interpreted this to mean all administrative costs and payment of the Sergeant-Major, a tall order for a fund relying on commisionaire's contributions alone, and the state of the fund needed close and frequent scrutiny, particularly during the early years.[14]

The Founder took no salary himself, relying on his own funds and on the unswerving support of his eldest brother, John Walter III. The financial state of the young Corps would not have been able to support such an expense but, in any case, he would never have accepted a situation where he became dependent on the contributions of his men. There was an urgent requirement for a source of revenue separate from the commissionaire's contributions which would both pay the running expenses and enable him to develop his Corps. It is not difficult to imagine his frustration at the first building he used for barracks. This could only accommodate a third of his men, when, in his judgement, many more than this needed constant supervision and could only be fully effective in employment terms if they were readily available around the clock.

Considerable hopes centred on a promising source of income identified in 1860 which appeared to meet the Founder's conditions as to its origin. This was the Corps' Circular which had evolved from the broadsheet designed to publicize the nature of the Corps and the current rates for hiring commissionaires. By early 1861 distribution arrangements had been developed. Commissionaires had already received several hundreds of pounds for delivering other circulars. The plan now was to produce a monthly *Commissionaires' Circular* with an edition of 20,000 copies. It was envisaged this would be distributed within a few hours, 15,000 copies in London, the remainder in other great cities, even some on the Continent. It would be free (giving commissionaires the responsibility of collecting money would have posed too many problems) and its income would derive from general advertising. It was hoped that public companies, institutions, charitable bodies, large hotels, and shopkeepers would find it advantageous to use. A single line was priced at 10/- a year. To ensure the strict separation of any income accruing from it, commissionaires were to be paid to distribute it as they would for other circulars.

In a letter to *The Times* of 12 February, 1861 (one of many published on his behalf since the Corps' foundation) Edward Walter appealed to its readers for their support. He said that working expenses had not exceeded £140 so far, clear testimony to his excellent husbandry and strict financial control, but now he needed to maintain an adequate staff for the growing organization and money to help build a proper barracks. The request was not for money (which he could not receive as the Corps was not a Charity) but for support through influential people taking the circular. The letter ended with the semi-warning that if he did not succeed in this request another plan would have to be tried within twelve months.

In the event success did not live up to expectations. Despite the proud motto at its head of 'Virtute et Industria', the paper lacked the vital advertisers. Letters from the Sergeant-Major to the Founder tended to take on a frantic note just prior to publication dates. The advertisers remained relatively few in number and of limited variety. One wonders how far the Founder's own strong feelings about what was acceptable for his Corps were responsible for a vetting system which involved a considerable proportion of rejections. In the early copies the advertisements were for such worthy causes as 'Homes for Destitute Boys', 'The Soldiers' Daughters' Home', 'Cambridge Asylum for Soldiers' Widows', 'The Couriers and Travelling Servants Society', 'Temperance Champagne' and 'King's College Hospital'. The advertisements were to increase in number although their services were still primarily directed towards ladies and gentlemen with military sympathies. Arguably one of the most interesting at this time was 'The Blakeley Ordnance Company' of Bear Lane, who were willing to supply cannon from 50 to 500 pounders to any interested party. There were also advertisements for Dentists, Mechanical Hair Brushers, the Caxton Steam Printing Press, Shrinkers of Trousers and Cigar Importers.[15]

A measure of the *Circular's* relative success was shown by its increased distribution to 35,000 copies, but by 1864 it had been reduced in frequency to a quarterly publication. The income could never have been large because the Corps Rules for that year talked of the Adjutant's salary being defrayed from either the profits of the circular or from the personal resources of the Commanding Officer. A wider range of advertisements might have achieved the Founder's financial aim but, more important, they would have been contrary to his own standards and the image he strove to create for the Corps.

At one time it seemed as if the foundation of a Corps Band would help to provide another significant source of income but it was to give the Founder infinitely more difficulties than the *Circular* and failure was acknowledged in 1864 when the rules firmly stated that 'All expenses connected with the Band will be paid by the Band Fund, and all Receipts arising from the engagements of the Band will go to the support of the men'. By making the bandsmen commissionaires any question of them subsidising the main body was dropped and any surplus in the Band Fund, after the musicians had shared out the engagement fees, was invested 'to form a retiring fund for those men who have remained in the Band for five years and upwards, and are unable to continue their profession'.[16]

The Band's main function was to publicize the Corps, to give it the dignity and style of a distinctive military-type organization at periodic parades and other Corps functions (eg funerals), and to project its image in the City of London in the parks and at various other public functions like the Lord Mayor's procession. During the initial stages serious problems were encountered. Endless discussion took place on the possible size and musical balance of the Band. While the musicians often failed to come up to standard musically, a large proportion were in debt on joining and were to remain so for a considerable time. The appointment of a suitable Bandmaster proved difficult and when the internal difficulties seemed largely resolved it encountered external opposition from influential quarters. This aspect will be discussed later but the Band utterly failed to provide any additional income for the Corps.

The continuing and seemingly intractable nature of the

financial difficulties led the Founder in 1864 to make a momentous decision. This was given full publicity in *The Times*' leader of 29 March. The leader opened with a panegyric of the Corps, its progress within the City of London and beyond, the basis of trust between commissionaires and the public, its strict but fair regulations, the admirable aim to find all good pensioners honest and sufficient employment, their good level of earnings, the moderate taxation to support the necessary expenses of the organization of the Corps and the commissionaires' welfare provisions. It then turned to the 'one feature of anxiety in this otherwise satisfactory picture. It will have been understood at once that the whole efficiency of the organization, the possibility of these men being qualified for the service they undertake, depends upon the maintenance of discipline of the Corps. That and that alone gives them the character which obtains the confidence of the public.' The nub of the discussion turned on the position of the Commanding Officer without whom, it maintained, the organization would collapse at once and the men 'become a mere mob without any order and respectability'. It noted that the Founder had discharged the office for five years at very great sacrifice of time and labour – at least eight hours daily. His services had been free and he was now obliged to look forward to handing over his work to someone else. To prepare for this the Founder and his friends intended to declare the Corps free of debt, despite about £1,000 which had been expended as an initial investment. However, to recruit someone of the right calibre he must be fairly remunerated. All this careful reasoning led up to an appeal for money, the very thing the Founder had tried to avoid through his first five years of the fiercest husbandry. Yet nothing less than the continuation of the Corps was at stake. If the Corps' structure depended on the position of its Commanding Officer and his subordinate officers ruling over individual, and often wayward, commissionaires with Olympian powers, they could never be dependent personally on funds contributed by those same commissionaires. If internal efforts to raise a higher (and separate) income had failed, and the Corps' future progress, if not its very existence, was in jeopardy, the money had to come from outside sources. *The Times*' proposal was for money to raise an endowment to provide not less than £350 a year for the permanent maintenance of the Commanding Officer and a small staff. It was to be a once for all public subscription and the paper emphasized the good value of supporting such an organization which enabled men to help themselves. Financial justification of the Corps' work followed. The readers were reminded that through it men were employed who had the greatest difficulty in finding work by other means and many were therefore prevented by the Corps from becoming chargeable to the country as in-pensioners of Chelsea Hospital. A calculation was made that if 250 Commissionaires were in-pensioners of Chelsea, instead of out-pensioners, they would cost the country, £12,500 a year instead of the current £4,500. This argument had obvious limitations. Chelsea Hospital was on a limited establishment only and it is unlikely it would ever have admitted the relatively young and healthy men accepted by the Corps of Commissionaires. Yet it formed part of a powerful and sympathetic appeal. Following the article was a list of the subscribers so far, with the amounts they had contributed along with the promise that all future amounts would be published every Monday in *The Times*.

The same edition contained Edward Walter's actual statement of intent. He affirmed that he would be happy to continue in his present office for a further two months only, at the expiry of which he intended to retire. His statement contained a list of ten provisional trustees to whom he would hand over the control of the Corps. These included Percy Wyndham, a prominent exponent of ex-soldiers' rights in the House of Commons, and General Sir J.Y. Scarlett, the Adjutant-General. At Edward Walter's request the latter had been tasked by the Commander-in-Chief, the Duke of Cambridge, to represent the Army in the transfer of control and the establishment of the Endowment Fund. The trustees were to be elected for three years, half of them being present or past members of the two services, their most important function being the appointment of the Corps' Commanding Officer, although they were also responsible for the general superintendence of Corps' affairs.

Five years, then, after the Corps' foundation, five years of endless planning and great commitment, we find the Founder seemingly changing his policy of strict financial self-sufficiency and declaring his intention to resign and to commit his organization into the hands of a Trustee Board. No original correspondence on this development survives; we have no more than *The Times*' articles and the Founder's printed declarations. The question remains open to how far the resignation was a ritualistic gesture to ensure the financial soundness of the Corps and to gain full commitment from the two services.

Guessing is a dangerous business but the specific declaration to retire after two months suggests it was somewhat more than a device, that he had reached a crisis in his five-year struggle, that he could, or would, not compromise over the rules he had forged with respect to commissionaires' individual conduct or to the Corps' financial structure. Despite the support of his family and its great newspaper, he knew that if the Corps was to grow in the way he had set, it needed fresh capital from outside, capital subscribed by officers and gentlefolk to help him in his work. The dignified and illustrious Trustees gave the Corps legitimacy and stature not enjoyed before. Their presence would doubtless help to make the appeal succeed and his resignation was on the table to be acted upon if the trustees did not give him their full support and true authority to manage the Corps' future affairs. After reading his personal letters one is in no doubt about Edward Walter's nerve and confidence; he must have known that, despite the restricted progress of his Corps, he had created a legend which had already furthered the cause of the ex-soldier in an unparalleled manner, that he could, if necessary, return to his general crusade regarding government posts for ex-servicemen with an enhanced personal reputation and much increased authority. He probably had no doubt in his own mind that the Trustees would hardly dare accept his offer of resignation.

CHAPTER FIVE

Progress – Within Strict Limits, 1864–1876

Any fear that Edward Walter would sever himself from the Corps of Commissionaires was dispelled by an announcement in *The Times* of 17 June, 1865, some fifteen months after the same newspaper had declared the need for the Corps to have a Board of Trustees to direct the Commanding Officer and a salaried staff to assist him. The newspaper reported the end of the provisional government. Henceforth, its affairs were still to be in the hands of a Board but one now permanently dedicated to the central aim of carrying out and perpetuating the fundamental principles and rules of the institution as devised by the Founder. This Board, made up in the main from senior serving or retired officers, acted to give the Corps a seal of continuance and stability which the Founder had long sought and for which he had requested the Duke of Cambridge's powerful assistance. To emphasize the importance and function of this body, *The Times* then gave details of the main Executive meeting for the year and the Report for the last twelve months, submitted to it by the Founder and Commanding Officer. Happily he was able to declare a startling numerical increase of some 20% together with an average rise in Commissionaires' wages of 5%. He was also able to demonstrate the role of the Corps (and himself) in helping soldiers and sailors find official posts. During the year Edward Walter had been called before a Treasury Committee assembled to examine matters connected with Government messengers. As a result of his evidence its members showed themselves fully satisfied that the employment of Commissionaires in such positions would be 'most advantageous to the Public Service'. In his Report the Commanding Officer went on to praise the work of his main salaried assistant, the Adjutant, a Mr Piper, and to offer his own services for another year, on the understanding that he should hold the appointment from the Governors. Any doubts, though, about the continued strength of his own position under the new control arrangements, or about his enthusiasm for continuing to lead the Corps, were surely dispelled by his confident and sweeping proposals for the future. The present barrack accommodation was roundly condemned as being insufficient and he revealed plans to establish a regular medical service and even a school for commissionaires' children. The newspaper then went on to describe the lavish compliments which the Board paid him for his work, and his humble response. This was to the effect that although he had hitherto exercised a despotic command he hoped it had been for the benefit of the men but that now he would 'only consent to receive his authority from the hands of the governors'.[1]

It is not hard to see the advantages of the new system compared with those early years when the Founder struggled alone. The Chairman, supported by his distinguished fellow Governors, was of senior military rank (not a retired Army captain). Together they could exert increased and more significant pressure on the two services to support the Corps financially. The Governors included 'soldier MPs' who could pursue, within the House of Commons, the subject of ex-soldiers' employment as commissionaires, and in a broader sense too, including their high suitability for certain public posts. Equally important, the criticisms levelled earlier at Edward Walter for his despotic and harsh rule were no longer possible. The final authority was now invested in a Board of most responsible men who reaffirmed the Commanding Officer's post from year to year and without whose support he could no longer continue with his duties.

The wisdom of establishing a permanent Board with its greater means of enlisting financial help was illustrated by the continuing slow progress in building up the Endowment Fund to support payment for the administrative officers. The fund was still disappointingly low. After an investment of £500 there apparently remained just £160 for expenditure in the following year. There was distinct hope, though, that the Board could generate larger and more regular responses, particularly from the two services. Yet once again it was the Walter family, through *The Times*, which set the example. It announced that from 1866 onwards interest from the balance of its Crimean War Fund would be allocated to the Endowment Fund, thus ensuring its viability.[2]

The aim of the Endowment Fund, viz to produce an income of £350 a year, was in sight and it was now hoped that the Founder would be able to unburden himself of the day-to-day minutiae of administration by delegating much of it to a permanent and adequately paid official. In the event this was more difficult than could possibly have been anticipated. The Adjutant, Mr Piper, was relieved by a Captain Stack, who rapidly resigned. The post was then held temporarily by Mr Haden who had acted as Secretary of the Army and Navy Pensioners' Employment Society. A Captain la Touche assumed the position in January, 1866, but fell far below Edward Walter's standards as far as dedication and exertion were concerned. He resigned after fifteen months. He was involved in a personal clash with the Commanding Officer, against whom he complained to the Executive Committee, but they refused to uphold his complaints. He was succeeded in a temporary capacity by a Mr Cain who was soon found unfit for such duties and whose appointment was terminated after seven months. Another ex-officer, a Captain Mainwaring, was appointed but he also resigned after about a year. Many of the problems arising between Edward Walter and his Adjutants at this time centred on finance, on which subject he was punctilious and demanding. Adjutants were personally held responsible for any inaccuracies in their accounts and for any damage sustained to Corps Property. In his opinion any such occurrences were the result of negligence on their part. The next incumbent, Captain Gilbey, held the post for about two years before being compelled to resign through ill health; for much of his term the Commanding Officer was absent abroad. Fortunately for both the Founder and the Corps, the next Adjutant, a Major Wilkinson, had the high standards, personal qualities and stamina required to work with Edward Walter. He was to earn the full confidence of his superior and to give the Corps no less thant eighteen years' loyal and unremitting service in this key administrative appointment.

One can appreciate the mixed reactions of some candidates to Edward Walter's method of determining their salary as Adjutant, one which reflected his individual notions of ensuring loyalty and application. These rested to a large degree on commercial principles and on self-help in particular, which he firmly believed led to the good of the organization. On appointment, Major Wilkinson's basic salary was £150 a year but there were certain considerable supplements. An allowance of 10/- a year was given for every commissionaire recruited above 300 men until the Corps reached 600. For every man above this target the allowance was to be 5/- a year. A commission of 10 per cent was available on all additional subscriptions and donations to the Endowment Fund (excluding annual subscriptions which had already been committed or those which should come from a special appeal). On the other hand the Adjutant was to be responsible for all losses that might be incurred by the General or other Funds, arising from his own neglect of the printed rules or written ideas of the Corps, or from the carelessness or dishonesty of those under his control. It was the latter clause which was so bitterly resented by some of the Adjutants before Major Wilkinson, for many ex-soldiers were inveterate 'fiddlers'.

Despite the potential liabilities, it was estimated that Major Wilkinson came to have an actual salary approaching £400, including commissions, no mean sum at this time, although his post was onerous and demanding. While the financial responsibilities might be open-ended, the incentives were considerable too.

Given the traditional carefulness of the Walter family, it might still seem a little surprising for the modern observer to learn how much emphasis the Founder placed on the strictest financial control at all times and in respect of the smallest amounts. While one might question the need to enlist so many resources to recover small individual debts from errant commissionaires this is to miss the main point. Doubtless his financial acumen was crucial for the success and continued growth of his Corps but he believed debts should be collected for moral reasons too. Individuals should never be allowed to get away with dishonest practices, for this affected the moral soundness of the Corps and could spoil the image for which he endlessly worked. Commissionaires had to be thought of as utterly honest and reliable purveyors and protectors of the public property. Edward Walter had also observed at first hand the errors and laxness which had played a major part in the financial difficulties experienced by the first employment organization which he had saved with his own capital. He saw clearly that further growth for the Corps would depend on its continued high reputation, including its sound financial practice. It would also need a wider financial base from which to pursue his ambitious expansionary plans. In its turn a larger Corps of Commissionaires would give him increased status, would help him to construct a more formidable pressure group and ultimately it would lead to greater support for the most important goal of all, soldiers and sailors being employed in the public Departments of State.

Fully confident about the rightness and importance of his cause, he now launched his next initiative for public funds. The immediate reason was a chance to purchase barrack buildings at an advantageous price. *The Times* mounted yet another campaign on his behalf.[3] A dignified appeal was made through a letter by Field-Marshal Sir John Burgoyne, an aged and revered figure who, the paper pointed out, had raised another important subject, the adequacy of the national defences, twenty-one years before.[4] In the accompanying leading article *The Times* drew parallels between his appeal on behalf of the Corps of Commissionaires and his earlier statements. It reiterated Edward Walter's main crusade, the advantage to the services and the state which came from opening up the lower offices of the Civil Departments to retired soldiers and seamen. This would be both economical and encourage good recruiting. While such measures were pending, though, the Corps of Commissionaires offered a good substitute. Men now had a 'department' of their own which could be enlarged to include all deserving ex-servicemen, men who might not be skilled or equal to the demands of manual labour but who had the special recommendations of courage, fidelity and trustworthiness. If the public supported the Corps it could, in Sir John Burgoyne's opinion, take up this challenge. The present roll of 377 men could be increased to 800 within a week if the public responded liberally to the appeal. If the public would subscribe the £5000 needed to buy a building for barracks, the rents paid by unmarried commissionaires living in them could go towards a larger endowment fund from which the additional officers needed to administer the increased numbers could be paid.

The logic of the article strikes one as rather laboured. *The Times*, at the behest of Edward Walter, was attempting to satisfy a number of requirements simultaneously. Through its appeal it once more referred to the need for ex- servicemen to be given employment in public posts; it alluded to the Corps as the body which could offer them employment in the interim and justifed the appeal on the grounds that an extension of its freehold would constitute a general extension of the Corps' administrative capability. Its final sentence aired yet another of the Founder's beliefs, namely that an extra 400 men were needed in the metropolis due to the limitations of the Metropolitan Police. He was now portraying ex-service Commissionaires as necessary and effective bastions of public order. This supported Edward Walter's action the previous year when he had offered the services of the Corps to HM Government in the event of trouble with the Fenians.[5]

Ten months later, when just £1,700 had been subscribed to the Building Fund, the Field-Marshal's letter was republished to kindle anew the public's awareness of the cause and to jog their consciences. Contributions came from four main sources: regimental subscriptions, individual serving and ex-officers, the general public and substantial individual benefactors. The Executive Committee had helped to bring about a higher level of contribution from the first two but the public at large remained relatively unmoved by the appeal. Fortunately for the Corps it was to enjoy invaluable support from one individual benefactor, the Duchess of Westminster. Further initiatives helped to raise the amount donated but in no way did they keep pace with the Founder's ambitions to extend the Corps' property holdings. The twelve years 1864–1876 were ones of rapid property acquisition in the metropolis. The means used to finance them were varied. For the most part the Founder acted on his own judgement to the partial bewilderment of the Executive Committee which in 1875 actually disclaimed responsibility over his latest proposed purchase. He went ahead with it nonetheless. The Committee minutes of the time recorded that they accepted, 'with thanks the offer of the Commanding Officer to take upon himself the contract into which he has entered for the purchase of 419 and 420 Strand, and disclaim all interest in the contract accordingly'.[6] Much of the time spent in Executive Committee meetings during this period was taken up with discussion about property purchase. The Barracks referred to in Sir John Burgoyne's appeal was purchased long before the capital requested had been subscribed. A loan from an organization called The Home Investment Society was taken out on behalf of the Officers' Endowment Fund. Other purchases were funded both from external credits and from loans taken out with the Corps' General and Savings Bank Funds.

It is difficult to reconcile, on the one hand, Edward Walter, the stern financial manager, insisting on the rapid payment of commissionaires' penny fines, even taking men to court for non-payment of their relatively small debts to the Corps, and the man who seemed almost fearless in his acquisition of property, borrowing heavily from outside sources and from other funds within his organization, brushing aside the cautious apprehensions of his own Executive Committee. The contradiction is less puzzling when one accepts that Edward Walter was very good with money. He had no doubt that property, more than any other medium, gave him the best opportunity to expand his organization. It was a respectable objective on which to launch an appeal. By obtaining it at an advantageous price it always represented an investment which could be realized if necessary. The income (partly through the full commercial rents paid by individual commissionaires) was above the average return. The property gave him new opportunties both to control and help his commissionaires and thus help to reduce the terrible wastage which severely hampered the Corps growth. It would also help add due weight and dignity to his organization.

In 1889, when his plans over property acquisition were well advanced, he summarised his main objectives here in a letter to intending benefactors.[7] His declared aims were as follows:- by acquiring valuable property to produce a net income for the Officers' Endowment Fund which exceeded 4 per cent (most safe investments at this time did not exceed an income of 3 per cent), and to provide barrack accommodation for a considerable number of single and married commissionaires together with the offices necessary to conduct Corps business. He emphasized the need to acquire still more houses so that a new barrack block could be built. Although he did not emphasize it in his communication, there is no doubt he saw the barracks as a formal club for his men where some could live and many more meet socially with their friends thereby developing a deep pride of identity with their Corps. Living in barracks, they were also readily available for immediate employment. By 1873 the Old Burgoyne barracks accommodated twenty-five men and had a band room, a recreation room, a kitchen and a laundry.

During the first building period, apart from the purchase of the Barracks (1-6 Exchange Court) in 1869, the 'Thatched

House Tavern' (8 Exchange Court) was acquired in 1870, Old Burgoyne and '46 Bedford Street' in 1872.

There were other initiatives in respect of property at this time. For over two years Walter had wanted to set up a Convalescent Home for his commissionaires where 'fresh country air, combined with good messing and quiet, might at least alleviate, if not cure, the sickness to which all are liable'. Accordingly a memorandum was sent out in February, 1876, to the two services requesting contributions towards a special fund, contributions coming from the non-commissioned officers and men of Her Majesty's naval and military services alone.[8]

In retrospect, this was the time of maximum fund-raising. After Edward Walter had decided in 1864 to put aside his first strict interpretation of self-help and ask for the extra funds needed to help the organization flourish, he seemed to cast his earlier diffidence away. With the Corps in the hands of an Executive Board, the Commanding Officer obviously could and would appeal with both energy and enthusiasm on its behalf. A number of different arguments were used. Apart from the need to pay adequately Officers of the Corps like the Adjutant, Assistant Adjutant and Quartermaster, he reverted, in 1872, to the question of the Commandant. The Executive Committee of that year was asked to consider raising some £10,000 to pay the Commandant's salary in the event of Edward Walter resigning and a successor being appointed. It was then decided to issue a further 'special' appeal for the Officers' Endowment Fund, the object being to raise about £7,000 to buy the Burgoyne Building, the rents from which would pay the future Commanding Officer's salary. The sum of £10,000 had obviously been considered too large, coming as it did with the many other objects needing financial support.

The amounts raised by 1876 were considerable by any standard. The Endowment Fund had received £14,432, admittedly almost half of which came from one wealthy and generous benefactor, the Duchess of Westminster.[9] The total amount expended by the fund on property was now £16,683 with further liabilities of £3,606. At today's prices (1986) this would be in the order of £1,936,000 and it is interesting to note that when the main block of properties was, in fact, sold in 1973 it raised just over £2,000,000.[10] Yet the pace was not allowed to slacken. Each Quarterly Circular contained a letter asking for further subscriptions to assist an Institution for which there seemed no limit in development and one of advantage not just to ex-soldiers and sailors but to the country at large. In the same year Lord Napier of Magdala, another distinguished soldier and long-time champion of the ex-serviceman, was persuaded to make an appeal in his own name for the Corps' Endowment Fund, duly published in *The Times* of 21 August, 1876, and supported by a hard-hitting leading article justifying the employment of ex-soldiers and sailors in the Civil Service. Yet another financial initiative undertaken was the request to employers for an annual subscription of 10/- on behalf of every commissionaire they employed, further indication of growing confidence in the calibre of the men supplied and the value of their service. The services were not allowed to waver in their support either. Apart from the appeal on behalf of the Convalescent Fund to non-commissioned officers and men, the Navy as such found itself sharply reminded that it should give more financial support.

One cannot explain fully these requests by citing the changed structure of the Corps. Plainly this was an important factor but behind the enthusiastic appeals lay a new-found confidence. The position was much changed from that during the first four years of the Corps' life. The Corps was now a success, and recognized as such. A complimentary reference made to it in the Parliament of 1867 was quickly cited by the Founder to prevent others from wearing uniforms akin to that of his own organization. The Corps now had a measurable portfolio of property, itself the tangible mark of progress in Victorian eyes. While invaluable support from the family newspaper continued, the organization was now officially in the hands of prominent men from service and civilian life who gave it both vindication and the practical benefits of their approval. Edward Walter himself shared in the success of his creation. The ex-Captain had achieved a measurable niche of his own in Victorian public life; he was an acknowledged activist for the ex-serviceman's right to minor public posts, giving evidence to successive committees on the subject. His most important single initiative, and the one on which so much of his success in other areas depended, was his Corps of Commissionaires. The daring experiment of transferring the stern and restrictive rules of regimental life in the unreformed Army to his own private force, a force dedicated to serve the commercial world of the 1870s, was now thriving. The ex-soldiers' reputation in civilian life would inevitably improve with further advances in its fortunes. He was also freer to consider and press forward with new policy initiatives. In the early stages he felt compelled to spell out in writing even the smallest task to his staff and to follow up their every action. Now he still wrote the most detailed instructions on matters of relatively minor policy – it was his style – but he could leave the Corps in the hands of a responsible official, its Adjutant (supported by an Assistant Adjutant). He did so in 1872 and part of 1873. During the winters of 1874 and 1875 he could do so with relative impunity when the excellent Major Wilkinson has assumed the Adjutant's seat. By regularly wintering abroad he was able to ease the strain on his constitution posed by his incessant labour and by the damp cold winters in Britain.

The most tangible measure of success during this time was seen in the increasing Corps roll. While one can dismiss *The Times'* exaggerated claim, contained in the appeal of 1868, that the Corps strength could be doubled in a week, the figure this implied was closely approached in 1877 and comfortably exceeded the year after. If this is thought to be slow progress one must never forget that it was achieved with no relaxing of the strict code imposed by the Founder. As he wrote in his Annual Report of 1872, 'The system of Fines which has existed since the foundation of the Corps makes it practically impossible for any badly-conducted man to remain in it . . . leniency, therefore, can be exercised only at the expense of those whose

Back of the Old Barracks in Exchange Court.

A single man's bedroom in the Old Barracks.

The ablutions in the Old Barracks.

character has made the Institution what it is.' The reward for upright conduct was the material reward of assured employment open to him as a member of the Corps.

By 1880 over 4000 men had joined the Corps of whom only 1883 had been considered good enough to be taken on the permanent register, the balance leaving before they qualified for it. Of those permanently registered 40% had featured in the defaulters book and 83% of defaulters were subsequently either dismissed or discharged the Corps or resigned prematurely. This gives weight to his comment that it was impossible for badly-conducted men to remain. The pattern up to 1880 was for only one man in four to remain in the Corps for any length of time. Yet this statistic tells little about the internal pattern. Those who were dismissed, discharged or expelled left relatively quickly, their average service being just over three months. On the other hand those who resigned, retired or died, served, on average, for five years nine months. The tendency, then, was for the core of reliable men to increase in proportion to the short-term commissionaires. This could only increase the degree of comradeship and sense of identity with their unique organization. In the following years many were to serve for a remarkably long time within its ranks.

Edward Walter continued to be disappointed by the number of dismissals and resignations through the seventies, although the bulk of these occurred with men newly joined. With his innate conservatism he partly explained it by thinking that the contemporary soldiers were worse than those of twenty years before. In 1872 the Corps only increased by three men, due to resignations rather than dismissals, but in this instance he held the staff partly responsible and his comments were withering.[11] His guard never slipped, his standards were as unequivocal as ever and they were applied to staff and Commissionaires alike. In 1876, writing to one of the Corps' subscribers, General Sir G. Hodge, he was sweepingly critical: 'The present race of soldiers are a deplorable lot in many points, a large proportion are drunkards and the officers themselves (to speak the truth) are much to blame.' One is left with the distinct feeling that he felt it would have been much better under the old regimental system.

As success was forthcoming in respect of his Corps so his position as an apologist on behalf of the ex-serviceman was enhanced. The two were of course part of his integrated campaign. We have touched on his efforts here before 1859; they in no way decreased subsequently. In the majority of his Annual Reports Edward Walter referred to the ex-serviceman's cause as a whole. In 1876, after commenting on the large number of early dismissals from the Corps, he averred that, 'A Soldier of the right stamp – for certain situations – not only ought to be, but is, far superior to anyone else who has not had the advantage of a military training.'[12] *The Times* published the Reports in full and they also featured prominently in other newspapers, particularly those with distinct military affiliations, like *The Army and Navy Gazette* and *The Broad Arrow*.

Exceptionally there was his book which brought together significant debates and articles on the employment of ex-soldiers and sailors in the Civil Service, including, of course, some of his own.[13] Two debates were selected. The first, on 29 June, 1868, was at the seat of Government, the House of Commons, where Sir Charles Russell sponsored a proposal to employ pensioners in subordinate positions in the public service. Sir Charles cited Commissionaires as men worthy of employing in such a way. He was supported by other members of the Lords and Commons, including Lord Hotham, who had presided over a Royal Commission on the subject in 1861. Unfortunately the Government of the day was not to be swayed so easily. W.E. Gladstone, on its behalf, gave a conditional reply agreeing to a Committee being formed, but he ended with the cautionary comment that it was not to be expected or desired that rapid progress should be made on the subject.

The second debate was held at the Royal United Services Institution on 15 February, 1871, where the Earl of Derby set forth a scheme to help recruit good men into the Army by holding out the prospect of permanent provision after service, in the form of civil employment under the state. During this debate, the soldier MP, Sir Charles Russell VC, a leading member of the Corps' Executive Committee, read a letter from Field-Marshal Sir John Burgoyne to Edward Walter on the subject. Prominent was Sir Charles Trevelyan who had been responsible for the reorganization of the Civil Service. During his address he specifically referred to the Corps of Commissionaires in terms which must have gladdened their Founder. 'This is on a small scale, but it is a good example of the class of men we have before our mind's eye. If the very moderate inducements which Captain Walter has been able to offer have been sufficient to form a corps of trustworthy men . . . how much more would a superior class of recruits be attracted to the Army if the public service were open to them!'[14] Strong support for the cause by the British Army in India was given by the Commander-in-Chief there, General Sir William Mansfield. A vote was taken which unanimously approved the motion, and it was agreed to set up a committee to 'call the attention of members of Parliament to this important subject with a view to its being submitted to the legislature at earliest date'. Members of the committee included two Walters, John Walter III MP and Captain Edward Walter, sure proof of his eldest brother's continued support and esteem.

The articles contained in his book, *State Reward for State Services*, including letters of Edward Walter, and others, to *The Times* and contributions from Sir Charles Trevelyan and Sir Charles Russell demonstrated the central position occupied by Edward Walter in the campaign for ex-servicemen to be given minor government posts, the distinguished circles he moved in and the contribution of the Corps of Commissionaires towards his central aim.[15] One can well imagine his mixed feelings over Sir Charles Trevelyan's remarks on the Corps. If there was to be no quick Government response, then pressure must be increased on it, the small-scale organization which Sir Charles referred to much expanded and its 'moderate inducements' improved too.

CHAPTER SIX

The Flagbearer, Recognition and Reward

In 1866 *The Times* had dramatically, if rashly, prophesied that the roll of 377 commissionaires could be increased within a week to 800, if the staff could only be expanded to administer them. This justification rested on there being at least 800 potential posts for such men in the metropolis. By 1876, largely due to the terribly high wastage of men newly joined, the manpower figure was still only 700, but the Founder, like any good commander, looked beyond a goal which was now plainly in sight. If London could only sustain some 800 or so, in what direction should expansion take? The answer appeared to be in the other great cities of the United Kingdom. In the earliest days provincial branches had been established but for want of local leadership and central assistance most soon withered away. In those days the directing staff had little or no time to spare for the provinces; they had difficulties enough in the battle to preserve the parent organization and money was far too short to subsidise out-stations.

By 1870, though, the Commandant felt confident enough in the position of the Corps and in its inevitable expansion, relatively slow as this might be, to look once more to the provinces as the logical area for growth. Such growth would also have the advantage of conferring a truly national character on the organization. The trend for moving beyond London was already apparent before there were branches to succour and encourage it. In that year he was able to tell the Executive

Committee that nearly a hundred men, some 20% of the total roll, were permanently employed in various provincial locations, while four years later he talked of an increase of 35% in this direction, a far better growth rate than for London itself.

From 1876 onwards the policy of creating or reviving the out-stations began in earnest. In 1877 branches were started in Belfast and Dublin with plans for a further one in the west of Ireland at Cork. They were retitled Divisions, a word which evoked a grander, more military image and one which denoted increased control from the central Army Commander. In 1878 a Division was founded at Nottingham. This made excellent progress under its honorary Commanding Officer, Major Poyntz, who, as Chief Constable, was an influential member of the local community. During the year the Adjutant made an official visit to the 'out-quarters' at Dublin and Belfast; they hardly yet warranted one from the Founder. The revival and expansion continued apace. In 1880 the Liverpool Division was re-established, after its abortive start in 1860, with Glasgow and Manchester following in 1881. The Manchester Division must have given the Founder particular pleasure. Previous to its reformation a group of pensioners, adopting the uniform and name of commissionaires, had operated there 'in a disorderly and ineffective manner'. By then Edward Walter had visited all the out-quarter Divisions and he personally opened the new one in Glasgow. In 1882 the Edinburgh Division was started and Leeds followed two years later.

The benefits of the greater support given by the Services to the Founder and his Executive Committee, together with the succession of initiatives taken to swell the Endowment Fund, were now becoming obvious. Initially the sub-divisions were expected to place a considerable financial and administrative load on their parent Corps. The Commandant estimated the joint cost of one commanding officer's and warrant officer's salary to be in the region of £200 a year. He saw no problem over the warrant officers' salaries which could be carried by the body of the Corps from the General Fund, but in his estimation the payment of the officers' salaries would require increased contributions to the Endowment Fund. It did not work out quite like this; some of the Divisions were started with honorary (unpaid) commanding officers, like Major Poyntz and these tended to attract their own financial support from local worthies. It was when the Founder moved to the appointment of paid commanding officers, whom he first trained at Headquarters, that the expenses became considerable. By 1885 payments of £190 a year had been made to the Commanding Officer of the Glasgow Divison, £150 to the Commanders at Manchester and Liverpool, £29 to Edinburgh and £53 to Leeds. On the credit side, though, employers and general subscribers in the sub-divisions were contributing just over £200 and the men employed in these locations now numbered almost 500.[1] This figure confirmed the growth potential now being tapped and these men, like all commissionaires, supported the General Fund. Other demarches yielded positive results. Advantage was taken of the Corps' 21st birthday to help extend the system of service contributions; in the case of the Army a new annual regimental subscription of £2 per Regiment was instituted.

Shortly afterwards, in 1882, a meeting at the RUSI under the presidency of the Commander-in-Chief, HRH the Duke of Cambridge, agreed on a special subscription for the Corps, the proceeds from the Regular Army going toward a testimonial for Edward Walter himself and those from the Auxiliary Forces to the Corps' Endowment Fund. The GOC Northern District made a personal appeal for this and raised the magnificent sum of £1100 for the Endowment Fund.[2] At the very least this ensured that the out-division which had been formed at Leeds would not constitute a financial drain on the Corps for a long time, if ever.

The struggle to extend the Corps, while not compromising the ideals of Edward Walter, had been both hard and protracted, but the years 1876–85 were ones of great achievement. In 1880 the Corps, now 1000 strong, attended divine service on Founder's Day in Westminster Abbey, thereby re-enacting that occasion 21 years earlier attended by the original nine men. By 1885 the Corps numbered 1567, a numerical advance in five years equivalent to that achieved in the first twelve. This in itself was only to be expected as the momentum gathered pace but the Corps now operated, albeit still thinly, over the whole country.

The permanent staff had grown in line with its increased responsibilities. In 1876 there was the Commanding Officer, Adjutant and Quartermaster (assisted by the Corps Sergeant-Major). By 1885 the Commanding Officer's responsibilities in the capital were delegated to an Officer Commanding the London Division, the Adjutant was aided by an Assistant Adjutant who was himself helped by a Deputy Assistant Adjutant (the 'Lady' Adjutant). By this time there were four paid Officers Commanding in the provinces.

With the increase in permanent staff came a concomitant rise in liabilities for them. In 1867 the cost of salaries and travelling expenses totalled £680; by 1885 this had almost trebled, totalling £1760. Fortunately income had risen accordingly; during 1885 it totalled £2481, of which £1173, or almost half, came from a combination of rental income and interest from *The Times* Crimean Fund. By 1887 the Founder estimated the Corps' assets, including property, to be in the region of £50,000 with a further £50,000 in the men's Savings Bank.

During the decade 1875–85 the Endowment Fund was cleared of all debts. This was a remarkable achievement and one not generally equalled by other voluntary bodies in this field. The National Association, funded in 1885, another body which flourishes today, whose scale of operations far surpassed those of the Corps, has never owned its Headquarter premises and its income from investment contributions has always been modest. Credit for the remarkable achievements in the area of Corps finance is due to a number of factors. Highly significant was the support and attendant publicity given by *The Times*, as also was the transfer of the interest accruing from its Crimean Fund. Important too were the efforts of the Executive Committee in raising higher and more regular contributions from the two services, while the internal funding, coming as it did from the compulsory subscriptions of commissionaires, increased as the Roll expanded and was subsequently boosted from subscriptions paid by their employers. Above all, the Corps owed so much in this respect to the Founder with his strict, accurate control of the day-to-day finances and his flair and daring with regard to the purchase of Corps property for the Endowment Fund. The 'Walter touch' never erred; he acquired several buildings, some at his personal risk, but in the main borrowing from the Savings and General Funds, themselves the fruits of his central financial policy.

Sweet as the progress of the Corps should have been to the Founder, sweeter still must have been the recognition and approbation of his work by the Regular Army. By 1880 his exertions were not only thought noteworthy and creditable but were considered quite the most important way of securing reserved places for ex-servicemen in the public departments. His advocacy was considered important not just for humanitarian reasons but for the good of the Army as a whole. If retiring soldiers could look forward to good positions in civilian life then better men would be encouraged to join, thus reducing the proportion of those poor quality men whom Walter firmly believed were well below the standards of the old Army. This was something he never ceased reminding senior officers about. In 1875, for instance, he made great play on the poor performance of ex-reservists as rural post messengers, conveniently ignoring the fact that these posts were ill-suited to grown men. The ex-soldiers themselves would not have been slow to show their displeasure at positions designed primarily for teenagers.

The Army was by no means niggardly in its appreciation of his work. There was the 1882 meeting at the RUSI which decided on a testimonial 'to express the gratitude of the officers of the Army and Navy to Captain Walter, the Founder of the Corps of Commissionaires, for his excellent services on behalf of discharged soldiers and sailors'. This resulted in the purchase of a most impressive silver centre piece which was decorated with flags and standards with two figures of commissionaires at the base. It included a handsome inscription of presentation. In October, 1983, its value was put at £16,500.

Also gratifying must have been the words of Major-General Cameron, the GOC Northern District:

'I consider the object in this instance is of such a national character that I am justified in treating it as an expectional case. . . . Should the Government ever undertake the duties of providing for discharged soldiers in Civil Departments of the State, I feel sure that the Corps of Commissionaires will be largely made use of to test the qualifications of such men previous to their employment . . . and will be carried on on a very much larger scale.'[3]

In May of the following year General Cameron also had a letter published in *The Times* which strongly justified one of the Corps' most criticized features, namely its iron régime. 'Its rules and discipline are strict (some say too much so) but all those well acquainted with our service must know that a highly disciplined probationary school is absolutely necessary through which to pass our soldiers into positions of trust if we are to secure the confidence of the public.'

Yet by far the most significant honour came in 1885 when the Founder was awarded a KCB. This recognition of his work by the establishment and the Royal Family was an honour which even his massively successful eldest brother did not share.

After so much endeavour, the years 1880–85 were a time for the Founder to savour. By this stage he was recognized as the leading apologist and authority on the areas of civilian employment which ex-servicemen could best undertake. Had he not been part of the 1877 Committee on the subject? Apart from the Army's testimonial and the award of the KCB, regular and generous tributes were paid to Walter and his work. The 21st and the 25th anniversaries of the Corps' Foundation were marked with particular regard. On 10 February, 1884, he addressed his men with utter confidence in the soundness of his organization and his self-appointed mission.

'The principles of good government, like those of true religion, are unmistakable and we may feel confident that the strict observance of the regulations which have contributed to our present prosperity will hereafter assume the same satisfactory results as those we commemorate this day.'[4]

The efficiency of the Corps had been sharpened over the years. It was in his judgement unequalled either within or outside the Armed Forces. In 1875 he had proudly stated: 'There was more discipline in this Corps than in any regiment and this was attained without any black hole or any other punishment for bad behaviour beyond that of dismissal.' He went on with the startling corollary that if the Regular Army could offer the opportunity of employment after military service it would have such a powerful hold on individual soldiers that there would be no need for the traditional range of punishments.

The process towards achieving such efficiency involved the combination of increased privileges on one side together with greater control on the other, the traditional military policy of carrot and stick.

Mention has already been made of the subscription scheme launched in 1876 for a Convalescent Hospital and Home for the Corps (an object never attained due to the limited response among NCOs of the Regular Army). This was soon followed, though, by the institution of an annual confidential report of 'the most searching nature' which was completed by every employer of a commissionaire. In 1878 Edward Walter extended his welfare provisions by appointing a Lady Adjutant to be responsible for all matters relating to the wives, widows or children of commissionaires. She also helped to increase the Founder's control. The rules were amended to state that, 'Every commissionaire joining the Corps as a married man must bring his wife and family (if in London) to see the Lady Adjutant within four days of his becoming a member of the institution. If marrying after admission, he must bring his wife to see the Lady Adjutant within a fortnight of marriage.'

In 1883 Walter instituted an Order of Merit based on the Army's long service and good conduct award. This not only rewarded good conduct but thrift and was not awarded until commissionaires had accumulated a certain sum in the Corps Savings Bank. The savings qualifications were as follows: for the badge £30, for the order of second class after fifteen years £60, for the order of the first class after twenty years £100. This strong encouragement to save, along with the other injunctions by the Commanding Officer on the subject, resulted in the average man's deposit increasing by 25% during the years 1880–85.

In accordance with the Founder's ideas the Corps had become a more military body. It was mustered in squadrons equivalent to units of the Regular Army. When on parade it was always supported by its own band; severe displeasure was expressed when, on one parade, incontinent commissionaires were compelled to fall out of the ranks. Should the need arise, Walter visualized his men as potential and quite substantial upholders of public order. 'We could muster amongst our able-bodied men a Battalion of Infantry, Squadron of Cavalry and Battery of Artillery, together with the usual complement of Engineers and other special branches of the Service in proportion, as able and willing as ever to do their duty should the honour and interests of the country require it.' One feels he would have loved the Corps to have been turned out to combat such an emergency.

By 1885 the Corps had taken on Walter's own high standards in accordance with the memory of his military experience in the 1840s. He was quite accurate when he compared its level of discipline as equivalent to that of the Army's select regiments. In fact he went beyond such standards; the Army's criteria for very good conduct certificates, for instance, were quite inadequate for him. To attain his objectives he required men who had qualities above the average. In 1880 he admitted that the average soldier was, and had been for many years, the bugbear of employers. 'He has not only lost the habit of work but is neither sober, industrious nor quick in adapting himself to the varied requirements of civil life. Consequently instead of having the call of the labour market, he is only engaged as a last resource.'[5] This was hardly suitable material for reserved entry to the Civil Service or other public departments. Although Edward Walter was now acknowledged by both Army and Court for his remarkable achievements with ex-servicemen he never had an exaggerated idea of the typical soldier. If anything his regard for him fell further. Around 1880 he appeared to be going through his own equivalent of William Shakespeare's black period when he came very close to despairing of human nature as exemplified by the rank and file. In that year he said it had taken '22 years and 4000 men of good character – according to War Office Standards – to make a thousand commissionaires'. Wryly he acknowledged the advantages too. He had succeeded partly as a 'result of natural selection, the best remaining; consequently the Commissionaire is sought for by everyone and his value is commensurate with the scarcity of supply.'[6]

Like other Walters, Edward was a stern critic of the Parliamentary system, holding that body primarily responsible for the 'regrettable situation', as he put it, both with regard to recruiting and outflow. 'If the service was *as it ought to be and might be*, the great mass, if not every individual soldier, might be as good and valuable as commissionaires. This, however, is never likely to be the case, especially under our miserable system of Government by Party'. His innate conservatism prevented him seeing the advantage accruing from Cardwell's Army reforms and the contemporary soldier whom he held in such scorn was the reservist. He compared the regard of regimental officers to the different quality of recruit to a consumer who faced the adulteration of the articles which he daily consumed. Edward Walter therefore placed special barriers against reservists joining his élite Corps. In addition to the normal caution and entrance fees, reservists were required, if single, to deposit £5, if married, £25, in the Savings Bank. If over 30 they had to deposit a further £1 for every year they exceeded it. No reservist was to be accepted unless in perfect health and of suitable physique. Reservists were to pay 4d extra every month to the Sick Fund from which they admittedly received larger payments if they fell sick. In 1889 he made further distinctions against Reservists, barring entrance to married men unless they had served a minimum of eight years with the colours and had saved £50.

However, by this time the majority of the outflow, from the Army particularly, were reservists rather than pensioners who had completed over twenty-one years' service. They also constituted a much enlarged outflow of men, men who, unlike the former middle-aged pensioners, needed genuine employment for the majority of their adult lives. Infantrymen, for instance, served on average for only seven years with the colours.

General Cameron, in his 1883 appeal on behalf of the Corps, clearly recognized the particular plight of the reservist. 'To find the discharged soldier a situation in private employ . . . is the greatest boon to him at a time when he is generally of all men the most helpless, and soon runs through his little bit of deferred pay, the fruits of his seven years' service.' Edward Walter, however, was far less disposed to acknowledge the reservist's premier place in late Victorian resettlement. It went against the image of the old Army he held so dear and he was convinced the Cardwell reforms, by creating a more rapid throughput, had led to a deterioration in both the moral and physical state of the Army. Both the Commander-in-Chief, HRH the Duke of Cambridge, and Edward Walter were at one in abhorring any change from the old Army of their youth and neither was prepared to minimize the practical problems posed for would-be employers by the new reservists. The Duke of Cambridge, in his position as Manager of the Royal Parks, said he could take an ex-serviceman who had retired on pension, 'but if you use the word "Reservists" and if it is on this class of men that your object is described, I am compelled to say I cannot take men (they must be permanently at their posts).'[7]

Fortunately for the reservist, by the late '80s and '90s the need to re-enlist in the event of small wars was much reduced. However, Edward Walter's reluctance to recruit them as commissionaires both hampered the growth of his Corps and opened the way for another organization to seize the intiative with regard to such men and, to a degree, for resettlement as a whole. Contrary to his convictions, by the mid '80s the quality of recruit was showing a marked improvement and a larger proportion of these men did have the ability to face up to the stern standards expected of commissionaires, even after their short-term engagements. They were also able to take on a wider range of civilian occupations where they were not discriminated against, and an employment organization catering specifically for such men, the vast bulk of the outflow, would in the nature of things expect to gain certain ascendancy over the Corps of Commissionaires within a measurable period of time.

The struggle for better resettlement still had far to go. The ex-soldier had little reason at this stage to feel he held an equal chance with his civilian counterpart. By 1880, though, Edward Walter had proved two things. One, that soldiers under his unyielding guidance could perform valuable tasks and earn the trust and respect of their employers. Two, once they accepted the high standards demanded, they had the stamina to perform their valuable roles for many years. The men, and there were many, who were discovered to be below standard were identified early. A good proportion of the commissionaires who retired or died whilst serving did so after twenty years of more.[8]

For these men the probationary school provided by the Corps to which General Cameron had alluded was a success. The public felt confident in the commissionaire operating both in London and increasingly throughout the provinces. They saw him as an exceptional man whose good conduct was guaranteed by his parent organization. This special band of brothers, though, could never embrace the bulk of the outflow and for a number of important reasons successive governments showed themselves unwilling to allot minor public posts to ex-servicemen.

Much could be made of the different attitudes held by continental nations towards their land armies compared with the United Kingdom. Yet while France and Germany unquestionably offered public posts to their long-serving men the continental systems relied very largely on conscription. They never offered such posts to their young short-serving men. By 1885 the bulk of retiring servicemen in this country were closer in length of service to the continental conscripts than to the regular soldier of the past. This helped to weaken Edward Walter's campaign for reserved posts and might have partly explained his attitude towards them. There were other influential factors. The major instance of posts being made available for ex-servicemen occurred in 1873 in the Post Office and the experiment was a dismal failure. While it is true many of those nominated for the posts by the War Office were not up to standard, the jobs of rural post messengers were not suitable for men with any ambition and no less than 103 of the original 220 nominees declined the appointments for this reason. However, in many quarters the example was seen as a demonstration of the unreliability and poor quality of the ex-servicemen for jobs in the public sector. It is regrettable that Edward Walter himself joined the clamour, taking the opportunity to criticize the inflated system of character gradings given to such men by the War Office.

It is also open to question whether the Corps of Commissionaires was the best vehicle to persuade governments to offer such posts. Commissionaires restored the public's faith in their old soldiers and sailors *when acting as commissionaires*. By being so selective, and so ruthless in discarding those not up to standard, Edward Walter could be said to have raised doubts in certain official quarters as to whether the average ex-serviceman could occupy minor public posts successfully. Like many reformers he was also fiercely independent and, unlike some, savagely critical of politicians as such. He wrote in 1884 that it was better for the interests of the Corps, 'that we should have no business relations with Government or any Government Bodies such as Corporations etc.' Referring to mayors and other dignitaries, he went as far as to say, 'I don't value their acquaintance or their custom .... We have never received the slightest assistance during the last 25 years from any official people in their official capacity.' Walter also maintained that the Corps was never a charity in any way but simply a business. He had no time for the weak or mealy-mouthed. 'The institution is intended for respectable men, not for beggars or improvident people. . . . [It is] not a club, not a charity in any sense or shape, and its members, though not rich, are well off and thoroughly independent.' He finished waspishly enough, trusting that his men were 'activated by as high principles as many of their richer fellow-countrymen'.[9]

By 1885 Walter seemed almost resigned to the fact that governments would never grant any considerable number of public posts to ex-servicemen. He had fought a concerted campaign for such posts through the pages of *The Times*, in his published articles, by means of his proxies both in the House of Commons and at the RUSI, and by his tireless advocacy with influential men in the Army and in public life. He had proved that the good ex-serviceman who met his high standards and who served within the ranks of the Commissionaires' Corps could do a useful job. Much had been achieved by one devoted and remarkable individual working in an area not accorded the priority given to other great issues of state. Yet at a time of undoubted success the nature of the battle was changing – and not in his favour. There was much for his organization and himself to do but no longer was he to have the centre of the stage. Senior officers in India were in the process of gathering influential support to establish an employment agency specifically designed to help the reservists, that group of men which Walter was slow to accept. If, in a greatly improved climate, it should succeed in offering ex-servicemen a range of occupations proportionate to both their ambitions and abilities, the need for reserved public posts and indeed for a Corps of Commissionaires would be likely to disappear.

CHAPTER SEVEN

A Rival is Born and Grows Apace, 1885–1899

The year 1885 marked a new era in employment opportunities for ex-soldiers. On 4 February a group of distinguished men, mostly senior officers, agreed to found a job-finding agency under the formidable title of an 'Association for Promoting the Employment in Civil Life of Reserve and Discharged Soldiers'. The prime mover was Colonel E.F. Chapman, RA, at the time Military Secretary to the Commander-in-Chief, India, Sir Donald Stewart, and later Quartermaster-General there. Edward Chapman convinced his chief about the serious problems being experienced by men passing into the reserve and thereby obtained a most useful ally. In 1884 Sir Donald wrote on his Military Secretary's behalf to many important men in the United Kingdom, including the Prince of Wales, the Duke of Cambridge and the Duke's distinguished soldier cousin, the Duke of Connaught. In a number of letters to the Duke of Cambridge Sir Donald referred to the ways in which Colonel Chapman proposed to help ex-soldiers, both practically and by enlisting general public support: 'Colonel Chapman has been engaged of late in getting Regiments in this country to work out some co-operation with the home Battalions, in view to the establishment of County registration offices under the military. . . . We want employers to recognize the fact that a man who has been a good soldier ought from his habits of order and discipline to be rather a better workman. . . . Unfortunately the leaning is rather the other way'.[1] He must have been somewhat disappointed by the conservative reply of the old Duke who showed his belief in laissez-faire and in the work being done by the Corps of Commissionaires. 'It is not a thing that can be done by the State; to be successful it must be supported by the employers of labour, great and small, throughout the country.'[2]

Despite the caution shown by the Duke of Cambridge, Chapman decided to go ahead with the new organization. Responsibility for bringing its sponsors together fell on Chapman's lawyer brother, Cecil, who from his London base worked tirelessly to this end throughout 1884. The men who assembled on 4 February already had a deep interest in the serviceman's welfare both during and after his service; they included Lord Wantage, General Sir Daniel Lysons and Field-Marshal Lord Napier of Magdala. Since 1872 the latter had acted as a trustee of the Corps of Commissionaires.

A number of decisions were made at that first meeting. A Central Office was set up to publicize the new organization which set out to meet the greatest contemporary employment problem, that of the reservist. 'No agencies exist for enabling the short-service soldier to learn what situations offer, and in what industries he has a chance of employment.'

There would follow 'some system of legislation which would serve as a medium of communication between employers and men seeking employment in all parts of the United Kingdom'. They were convinced that soldiers re-entering civilian life were good material and sought to make sure of it by limiting assistance to soldiers who, while serving, had character gradings of good or above. Unlike the Corps of Commissionaires they required no further guarantees and, unlike it, fees were not to be charged against either employers or intending employees, although like it the Association was to be in no sense a charitable organization. The Association quickly accepted for itself that responsibility first undertaken by Edward Walter of promoting the employment of soldiers in the wider sense. It urged on both the public and the Government the right of the soldier to civilian employment after good service with the colours and sought to bring to the notice of Parliament 'the duty of the State in its several departments with regard to this matter'.[3]

In a circular written in October, 1884, at Simla Edward Chapman referred to the greatest source of apprehension felt by reservists' would-be employers, their being called to the colours at short notice. He even raised the need for 'sufficient and satisfactory compensation, under National authority to employers of labour, whenever, in case of National emergency, they are forcibly deprived of the services of men who belong to the Reserve Forces'.[4] Without this, he warned, the privilege of recruiting by voluntary enlistment could be lost.

From the outset the new organization represented a serious rival to the Corps of Commissionaires. It aimed itself at an important field which the Corps showed a reluctance to enter. It was designed to work directly with the Regular Army, with those serving officers at Regimental District Headquarters who were responsible for the newly-established registers of men who needed civilian employment. It quickly captured the imagination and loyalty of powerful supporters who had been well briefed by the personable and able Chapman brothers. In December, 1884, no less a person than the Secretary of State for War, Lord Hartington, gave his strong endorsement and even questioned the Duke of Cambridge about his own lack of enthusiasm for it.[5]

What do we know of Edward Walter's reaction to the new body? Evidence is limited but Edward Chapman certainly wrote to him from India about his proposed scheme. It was a careful letter which took pains to acknowledge the achievements of the Corps of Commissionaires. Two subsequent letters, both from Cecil Chapman, would suggest that Edward Walter replied fairly noncommittally but indicated that he would be glad to meet Edward Chapman to discuss his plans.[6] Cecil wrote to say that his brother was not returning to England until March, 1885, but that he would be glad to discuss the proposed organization. A second letter mentioned that a Colonel Cameron, working on behalf of Edward, would very much like to see Edward Walter and would be glad to meet him at his club. 'He has an idea that you would disapprove of the scheme as a sort of interference with your own but I pointed out to him that we were likely to deal with very inferior material, that we could never expect to establish anything like your system and that even if such an idea could have arisen, rivalry was simply out of the question. He went as far as to say that your approval of the scheme would make all the difference in the view which he would impress upon the War Office.'

Despite the tribute paid, one can imagine Edward Walter's thoughts about any organization which needed to rely on War Office support and how far removed this was from his own completely independent body. The best indication we have of his attitude to the National Association comes through a letter he wrote in 1889 to the Officer Commanding the Belfast Division of the Corps.[7] This division was obviously concerned about the effects the Association could have on it. Edward Walter was uncompromising: 'I do not at all agree with Mr McNeil about the possible damage the Reservist Society may do us. I have hundreds of times expressed a totally opposite opinion and have written officially and strongly in the opposite sense. Even if I entertained different views, *I should not be such a fool as to express them*. I consider the Society a "Godsend" to us. When an inferior man comes to us we can now pass him on to the Reservist Society who practically get our "leavings". . . . We shall always get the best men in the Army. It is absolutely impossible for anyone to damage the Corps, except those who are unworthy members of it, or who neglected their duties.'

To Edward Walter the Corps of Commissionaires now represented true resettlement; other alternatives or measures of collaboration were rejected outright. 'I wish the War Office joy of their institution. They were very angry when I refused to join them, and much displeased when I told them that in place of being afraid of them, I would prefer giving them an annual subsidy out of the Funds of the Corps to seeing them broken up.' Such a statement would have made few friends in either the War Office or the National Association and his concluding remarks showed that his antipathy to the Reservist was as strong as ever. 'It is ridiculous to compare a "Pensioner" or "Reservist" with a Commissionaire. The latter is well worth 20 per cent more in the labour market. . . . When the employers have had a little experience of the class of men now admitted into the Army, and are turned out of it in a few years with a *very bad education*, they will not patronize the "Reservists Society".' He went on to tell Major Boyton, at Belfast, that an officer in charge of an official register would 'usually be only too glad to send his best men to you'.

Time was to show who was right about whether the Reservist was such poor material in civilian labour terms and how far potential employers favoured commissionaires, but henceforth the National Association and the Corps were to move on separate paths and the opportunity for a more co-ordinated service had been missed. Ten years after his letter to Major Boyton, Edward Walter was approached by Sir Lumley Graham of the National Association on their working together at Leeds. The proposal was rebuffed. With hindsight it is easy to criticize Edward Walter's inability to recognize the virtues of the Reservist and it is also relatively easy to talk of some natural form of co-operation between the two bodies. The fact was that the nature of the Corps precluded co-operation if it meant a lowering of standards or any loss of independence, the twin pillars of its existence.

The years 1885–1899 saw a steady growth in the size of the military establishment. This, together with the short-service system which caused a dramatic rise in the annual outflow of servicemen, put a great strain on a recruiting system dependent on volunteers. Whilst the numbers of men willing to enter the two services during the late '80s and early '90s were barely sufficient, in the main the position was never desperate until the very end of the century. One obvious way of alleviating the problem was by recruiting a better stamp of man who would have a lower incidence of invaliding or discharge for misconduct, but this was not easy to achieve due to the continuing suspicion of military life among civilians. There were two other more fundamental solutions. One was to revert to a longer-service Army. This solution was favoured by many, including the old Duke of Cambridge, W.H.Smith, Secretary of State for War from 1886–92, and Field-Marshal Lord Roberts, the last Commander-in-Chief, who was strongly abetted by Edward Walter. The other solution advocated, when a reversion to a long-service Army was no longer seen as practicable, was to adopt the continental system of recruiting by conscription. This movement, however, did not gather momentum until the early years of the twentieth century, after Edward Walter's death. It was led by Lord Roberts who, unable to achieve a longer-service Army, came to see some form of conscription as inevitable.

From its foundation the National Association accepted the current system of recruitment. Indeed its very existence depended on there being reservists. Its approach was to work for better civilian opportunities on their behalf, which, in turn, would lead to the recruitment of an improved class of man. As the Chairman at the first annual meeting put it, 'No one can overstate the importance of having a contented Army – an Army of sufficient strength to guard and defend our Empire – an Army cheerful in its present service and confident regarding the future of its members, when they shall have passed from the service to be absorbed in the civil population.'[8] These annual meetings became brilliant public relations exercises where successive Army leaders, including the Duke of Cambridge and Secretaries of State for War, were called upon to speak. Not surprisingly at such gatherings they publicly pledged their support for the Association. At the inaugural meeting the Chairman also found time to deplore the lack of action taken by Parliament in offering minor government posts to servicemen and then demonstrated the basic differences which existed between the new body and the first two in the field. The Association pledged itself to gain official support in obliterating the line which, it maintained, had separated the soldier from the civilian. This would not be through total state provision, though, for 'State provision weakens the energies of people and, left to their own exertions, they become self-reliant, hardy and enterprising'. While acknowledging the successes achieved by the Corps of Commissionaires, he envisaged the solution to lie in persuading the Government to devote a fair share of the appointments under its control to servicemen (in the main reservists), while encouraging the Association in its endeavours to promote honest employment for ex-soldiers. Whereas Edward Walter was fiercely independent and created a special form of employment primarily for pensioners, the National Association saw itself as a voluntary organization which would welcome official support and indeed relied on War Office approbation from the outset. The employment it sought was for young men through normal civilian outlets, not through a private uniformed body, and at a higher level than the placements achieved by the oldest employment-finding body, the Pensioners Employment Society.[9] While it appealed strongly for more government posts to be made available, unlike Edward Walter it did not see the reservation of such appointments as the root solution to resettlement.

From the outset the Association promoted itself as the best means of providing orthodox employment for ex-servicemen. The late Victorian period, with its continual employment difficulties, was hardly a time for widespread allocation of civilian posts for ex-servicemen. To that extent Edward Walter identified himself with an unattainable objective and the National Association's contemporary stance must have seemed more realistic and more reasonable to political and military leaders alike. By its very compromise it served to lessen his chances of success and to wrest such initiatives to be made in this direction towards the National Association.

The outcome of the contrasting objectives adopted by the Corps of Commissionaires and the National Association can be seen to some extent in their different growth rates. During the years 1885–1898 the Corps grew steadily but slowly. The roll of 1567 commissionaires in 1885 rose to 1835 in 1890 and 2464 in 1898. In thirteen years the Corps had increased by 893 men or by under seventy men each year. In this same period the total number of men joining it was 4727, or 363 annually. While the majority of men entering the Corps were pensioners with a maximum of 20-25 years service before them, the remarkable statitics of five men being needed for every single unit increase in the Corps' roll arose for a different reason. This was the failure of new entrants to come up to standard: 100 men were dismissed, expelled or deserted during 1885, and the annual total of dismissals remained around this figure until 1892 when it dropped by about a third to around 3% of the total roll. Others left voluntarily, usually after a relatively short time. In 1885 these totalled 152 and the annual totals remained close to this figure until 1898, when 229 men resigned. The average over the period was no less than 8% of the roll. Two other significant forms of wastage occurred, but they were quite small compared with the first two. These were through discharges and death. Men were discharged because of unfitness to carry out their duties, normally through ill health or age, but these averaged about 25 annually, certainly under 2% of the total. Deaths were even lower, averaging about 1.5%. Looking at the statistics in another way, during 1885 almost 19% of the total roll were casualties from the Corps, a remarkably high figure; by 1890 it had fallen to just over 16% and in 1898 it was below 14%. This represented a reduction of over 25% in the annual wastage during the period but it still acted as a major brake on the Corps' growth. In contrast once new entrants accepted the system they tended to stay for a remarkably long time. The average length of service before 1900 of commissionaires from the infantry on the permanent roll was 24.9 years and from the 60th Rifles 26 years.[10]

Other conclusions may be drawn from these figures: by 1898 the Corps of Commissionaires catered for an élite minority who, once settled, continued in its ranks for a long time. But its standards and levels of regimentation were unacceptable to many men who joined. Such men tended to leave relatively quickly, but this wastage, although still high, had dropped by the end of the period. Contrary to Edward Walter's feelings, the calibre of the ex-soldier seemed to be rising (he certainly made no attempt to lower his own standards) and this was not entirely divorced from the relatively small but slowly increasing numbers of reservists in its ranks.

What such figures fail to reveal is the other main brake on growth, namely finance. In the first five years Edward Walter struggled on alone, supported by his family and friends, until he decided that the Corps had no real future without setting up a fund from which to pay officers of the required calibre. Even then progress was painfully slow. For long financial support from the two services was somewhat grudgingly given and less than might reasonably have been expected. Cecil Chapman, in his letter to Walter on the formation of the National Association, paid tribute to his progress 'in spite of opposition and the entire absence of support from people in authority'. This was the penalty paid by a man ahead of his time and one who, particularly in the early stages, did not enjoy the support of many influential contacts. In this respect his failure in 1859 and in 1864 to achieve 'Royal' status for the Corps must have been a heavy blow. The reasons put forward by Sir Charles Phipps of the Royal Household for not granting the title in 1859 hinged on the independence issue and Walter's own relative obscurity. '(Her Majesty) does not think that it would do for Her Majesty to give to what is after all a private undertaking the Title of "Royal", which if once given must be attached to it permanently. . . . Under voluntary Superintendence there is no security for its permanence, good management or even respectability.'[11] It was the same Sir Charles Phipps who blocked the second application in 1864. A further application in 1899 still failed to achieve Royal status for the Corps.

There is no doubt the Corps required powerful external support. In its case growth was bound to be expensive, much more expensive, for instance, than any comparable expansion on the part of an Employment Agency like the National Association. The latter operated from rented premises, had a paid Secretary, but incurred relatively little expense for its branch offices. If positioned in Regular Army Barracks they were normally free of rent, if in rented accommodation they relied on local subscriptions to defray the cost. The Corps needed to maintain that degree of discipline and supervision intrinsic to its role through a body of paid permanent staff which the Founder estimated needed to be increased by a further Staff Officer for every 400 men. In the case of its most promising area of growth in the UK, the outstations, extra expense was inevitable and prolonged. Only after 10–15 years could they be expected to become self-supporting. In 1898, for instance, the excess of expenditure over income for the salaries and expenses of staff in the outstations was £619.13.7, all of which had to be met from the Officers' Endowment Fund with its average return on investments running at about 3½%.[12] Between 1885 and 1898 only three outstations, Birmingham, Newcastle and Wales, were added to the existing eleven, although these helped to fill major gaps in the countrywide service offered.

In contrast the National Association and its affiliated Regimental Agencies placed 174 men by the end of its first year, 1462 during 1889 (after a streamlining of the Official Regimental Agencies into County Soldiers' Employment Associations), 3886 during 1893 and 5554 men during 1898.

Figures alone tell less than the whole story. The Corps of Commissionaires took on immeasurably more responsibilities for its men and it had blazed the trail. Yet the short-service system did result in a massively increased outflow; in 1896 it was 24,051 of whom 18,246 had reliable characters. The National Association expressly set itself to assist the men with six and nine years' service, who constituted the vast majority. It clearly had some way to go in achieving its aim but Edward Walter's Corps, its eyes set predominantly on the pensioner and with its rate of progress severely restricted, seemed destined to occupy a branch line in resettlement terms even if its standard of service remained unequalled. It had made a fundamental contribution towards a more favourable employment climate for ex-servicemen, but it could hardly continue to compete in numerical terms. Despite the relatively modest posts it sought, the original employment organization which

Edward Walter revived, the Army and Navy Pensioners' Society, had also succeeded in raising its annual totals sharply compared with the Corps. In 1896 it placed no less than 2353 men. Without doubt the impressive momentum generated by the National Association owed much to its self-appointed role as *the* official employment-finding agency for servicemen and to central support, something not at all to Edward Walter's liking.

It shared, or was granted outright, military accommodation and clerical support; it received an annual War Office Grant of £200, raised to £500 in 1893 (but reduced soon after) and it came to be seen as the major body in the field which attempted – with only mixed success – to obtain a better class of employment for its registrants. It also took upon itself a version of Walter's own campaign for alloting more minor government posts to ex-servicemen. By 1899 in many respects the rival had supplanted its distinguished forerunner.

CHAPTER EIGHT

The Australian Venture: An Age Draws to a Close

Edward Walter was 61 years of age when the National Association was founded. His early struggles were behind him; he had recently been honoured by both the Queen and her two services and the role of his unique organization was acknowledged throughout the country. One might have expected him now to leave the heat of the continuing campaign to others. Yet there remained a further dream he had long determined to pursue. He often said there was no limit to the growth of his Corps given the right men and the necessary infrastructure and he foresaw it operating not only throughout the United Kingdom but across the Empire as well. Accordingly, during the annual inspection in 1886 he announced his intention of sending a group of thirteen hand-picked commissionaires to Australia under Major Hawkins, Commanding Officer of the Edinburgh and Glasgow Divisions, who had volunteered his services. This was to be the spearhead which, if successful, would presage other divisions in New Zealand, Canada and South Africa. It entailed sacrifices, for this commitment caused the establishment of new outstations in the United Kingdom to be postponed.

The detachment arrived in Australia in the Spring of 1884 and Major Hawkins and the Founder took the P & O liner *Victoria* in the following October. Leaving Major Hawkins Sydney-bound, Edward Walter and his family disembarked at Bombay. From here he travelled over 5000 miles throughout India visiting military units and exploring the possibility of recruiting men who had completed twelve years with the colours. The individuals he sought for Australia were full-term reservists rather than pensioners, men who were physically better equipped and available in large numbers for the expansion he confidently expected. They would provide 'a constant supply of able-bodied and well-seasoned men, who, having completed their full service of twelve years will prefer to settle for the various colonies with the certainty of constant work at good wages in place of returning to a country where the Labour market is overstocked.' His imagination soared; he looked to establishing a second Division at Melbourne and yet more in Australasia so that by the next October (1889) an employer wanting a commissionaire 'in any part where no Division exists' would be able to get one from Sydney.

The selection of reservists also gave Edward Walter the opportunity to revive another of his long-held ambitions, of using the Corps as a military force in times of danger. By offering reserve pay to commissionaires stationed in Australia their services could be at the disposal of the Commander-in-Chief, India, in the event of an attack upon Afghanistan. They would be within twenty days of Quetta and would not interfere with any reinforcements from Europe.

One can measure Edward Walter's enthusiasm for the Australian venture by the high level of financial commitment. He put the expenses at £1000 for the initial two years, virtually the same as the cost of all the United Kingdom out-divisions over the same period. Tragically his greatest single venture was not to succeed, for two incalculable but crippling circumstances. One was a personal tragedy, the other Australian cussedness. The loss of his daughter Florence, who died of cholera in the Bay of Bengal on 10 February, 1888, forced him to cancel his plans to join Major Hawkins in Australia. The other occurrence came from no sudden blow but it was even more serious to the success of the initiative.

At first all went well. The original thirteen men were immediately employed at excellent wages (75% above the home level) but in December, 1888, the *Sydney Echo* reported a meeting of the Trades and Labour Council where disgust was expressed at the prospect of more commissionaires arriving to supplement the number already there.[1] Another meeting deplored their employment to the detriment of local labour on the grounds that compulsory subscriptions required by the Corps resulted in them working for considerably lower rates of pay than the locals. In January the Labour Council further considered the Corps in Australia and its main concern now appeared to be concentrated on the treatment of individual commissionaires, not only because of the necessary subscriptions but the stringent selection involved. 200 men had been enrolled but the actual Corps' strength stood at no more than thirty. The Council was also alarmed that some commissionaires had been employed by the Australian Government. These and other charges were answered in the press by Major Hawkins and his Sergeant-Major. Yet a public meeting was now convened by the Labour Council during which it was addressed by Labour MPs and several ex-members of the Corps. This condemned the organization 'as a serious encroachment on the rights of the labouring classes'. By April, 1889, the *Sydney Morning Herald* reported that a cablegram from Major Hawkins had informed Captain Walter that 'the Labour Unions will probably drive the Corps of Commissionaires out of Sydney'.[2] Not all Sydney opinion was hostile, though; several articles supporting the Corps appeared in a number of newspapers but considerable damage had been done. For instance, after the public meeting no more men were employed in Government positions.

Back in England, at the annual inspection of the Corps in June, 1889, Edward Walter still harboured and publicly expressed his hopes for the Australian venture. Yet the adverse pressure was being maintained in Sydney; in May the Labour Council sent a deputation to the Prime Minister and a representative to other parts of Australia in order to drum up opinion against the Corps. It attempted to establish a boycott of commissionaires' employers in Sydney. In response to the deputation the Prime Minister agreed to set up an inquiry. This reported that commissionaires in public offices 'were more obedient and reliable than many officials who have occupied similar posts in the past'. As a result the Government decided not to interfere. Another challenge appeared in the form of the 'Australian Commissionaires Guarantee Society' set up in direct competition to the Australian Corps.

The cumulative effect of the opposition succeeded in making the climate too adverse for the Corps to progress in Sydney and the Corps' Executive Committee Report of 1889 regretted 'that the branch at Sydney, Australia, had not proved a success. On the other hand the out-quarter Divisions in this country such as those at Leeds and Manchester are in a flourishing condition.' Edward Walter's own report to the Executive Committee for 1889 was hardly more forthcoming: 'I greatly regret that our want of success at Sydney has compelled me to break up the Division there, and bring back Major Hawkins.'

Despite the terse references – failures rarely warrant undue publicity by those directly concerned – this was a grave setback. The ambitious plans for spreading the Corps across the Dominions had suffered a most serious check at the hands of a caucus of bloody-minded but determined trade unionists in Sydney. Funds were not available to risk other failures of this kind and a proposal in 1888 from Vancouver to establish a branch there met with the greatest caution at Head Office. Despite isolated initiatives, overseas branches of the Corps of Commissionaires were not firmly established until after the First World War and their story belongs to a later stage.

Once more Edward Walter was ahead of his time in his long-held vision of the Corps operating throughout the Dominions. By the late '80s, however, it was not an isolated notion. The new appreciation of the Empire in late Victorian England led other crusading organizations to look for expansion beyond the UK. For instance, after 1887 another uniformed body, the Boys' Brigade, just four years after its foundation, witnessed its work spreading outside Britain, so it was said, 'with all the violence of an epidemic' while in the early 20th century the Boy Scout movement virtually started with Dominion involvement in mind.[3] In one instance, the Naval and Military Emigration

League, an organization was designed specially to help ex-servicemen re-establish themselves overseas.

Apparently almost without another look overseas, emphasis in the Corps of Commissionaires returned to extending the outstations in the UK. Yet the cost of the setback had been considerable, if somewhat less than might have been expected. In 1889 Sydney cost £250 and it was because of this that Edward Walter saw no prospect 'of adding to any of our out-quarter Divisions at home'. Plans for divisions at Birmingham, Bristol and Newcastle had to be postponed.

The mood of the age was not favourable for another major new appeal but nothing less could bring about a new surge in growth as the traditional sources were fully worked over. By 1888 the Army was contributing most loyally: all regiments, except three, were subscribing and in that year General Roberts collected £400 from the Royal Artillery alone. It was thought that some increased contributions could still be obtained from the Navy but all civilian employers now subscribed annually on behalf of their commissionaires. While *The Times* could still be relied on to publish the Corps' activities, the one-time Thunderer was beset by new competitors and financial problems of its own. Its great editor Delane, who gave the Corps particular support, had gone over a decade before and the hour glass was fast running out for John Walter himself. In any case the achievements of the National Association, working as it did from a much smaller budget, made any major initiative for a relatively asset-packed organization somewhat inappropriate. The 1896 Founder's Report regretted the lost chances caused by financial restrictions but partly consoled itself by the inevitability of the Corps' growth: 'We cannot fail to grow slowly and at any rate we are able to keep out of debt.' The note of resignation here could have been caused in part by the lack of public response to Walter's letter in *The Times* which explained the establishment of a new Division at Newcastle and the need to support it. Another telling factor acted against more rapid expansion at this time. Jobs were difficult to obtain, especially permanent ones. From

The bar in the Old Barracks.

Playing billiards and chess in the Old Barracks.

Taking a commission outside Exchange Court.

the early '90s a general depression in trade affected the Corps and caused progress, particularly in the out-divisions, to be slower than expected. Ironically it began to lift just before the upheaval brought about by the Boer War.

The National Association was itself not immune from such conditions but they effectively prevented the Corps from reducing at all the great numerical imbalance in the different kinds of employment opportunities offered by the two bodies. The Founder and his Corps had now to settle for a rate of growth within the present funding arrangements although some bonuses could be expected from individual benefactors from time to time. It could never go hell-bent for growth anyway. Internal restrictions came from two measures fundamental to its character: the continuing need for careful and efficient supervision and firm discipline. In respect of the Endowment Fund for officers' salaries, it was the Founder's unshakeable conviction this should never be swelled, even temporarily, from alternative sources which depended wholly, or in part, upon, commissionaires' subscriptions. The continuing emphasis on a rigid disciplinary code caused considerable capital to be required for improved accommodation. At Headquarters this was for about 200 single commissionaires and better barracks were added at Birmingham, Leeds and Glasgow. The work at Headquarters was so extensive it took about nine years to complete.

Some promise of accelerated growth lay in Edward Walter's eventual recognition of the increased proportion of reservists in relation to pensioners' who were coming out of the Army and their growing significance to his Corps. He sought full-term reservists in India but he would never compete openly with the National Association which went directly for reservists, many of whom were too immature to reach his high standards. Essentially to him the Corps' special value lay 'in the quality, not the quantity of men who compose its ranks'. Yet he knew numbers had a significance too and came to see a definite place in it for some reservists if they agreed to make sensible financial provision through the Corps' savings scheme for their eventual retirement. Oblique credit to reservists as well as pensioners came in his tribute to 'the progressive improvement which has taken place in the position and education of those who have been admitted during later years'. Unfortunately the stern financial conditions he continued to stipulate before admitting reservists prevented many good men from joining the Corps. It is hard to see why he did not ameliorate these conditions although the late twentieth century view is significantly different from that of a century before when short service still had its powerful opponents and when a number of attempts were yet to be made to change the terms of engagement back to the longer service Army of before the Cardwell reforms.

While by the early '90s the Corps had been forced to relinquish its earlier leadership of the campaign for special employment privileges and its numerical pre-eminence had gone for ever, the quality of its men, its functional efficiency and its reputation as an organization had never been higher. Although the refurbishment and extension of the Headquarters accommodation was not completed until 1898, the service provided from the old, mainly timbered, building in the previous two decades had become a byword throughout the metropolis. The Corps had established itself as a very important and useful organization which had become part of the London scene.

In this context it was only to be expected that a man who relied on the prompt and secure transmission of information like Sherlock Holmes, a stickler for detail, should make considerable use of the commissionairing service when in London. In 'The Case of the Blue Carbuncle' Holmes used a commissionaire by the name of Peterson who on one occasion was instructed to insert an advertisement in all the London evening papers. At another time Holmes had to call into a District Messenger Office in Regent Street to engage a Corps messenger, while in 'A Study in Scarlet' an ex-Sergeant of Marines arrived at Baker Street with a message from Inspector Tobias Gregson while the great detective and his friend were still at breakfast. The relatively early hour of this delivery indicates a service which operated around the clock. The commissionaire was in civilian clothes because his 'uniform was away for repairs' (although Holmes inevitably identified him) but he was swift, conscientious and respectfully dignified, clicking his heels together and raising his hand in salute as he left. Holmes called him 'a man with some amount of self importance and a certain air of command. . . . A steady respectable middle-aged man too.'

Hardly surprisingly, Rudyard Kipling, with his knowledge of the soldier and his empathy for him, when serving or retired, was also fully familiar with the Corps and the role it played. In his inimitable style he was not afraid to comment in ironic vein.[4] He described an ex-Troop-Sergeant-Major's wife charing while he was a commissionaire:

> . . . .'the pay-bills to better,
> So if me you be'old
> In the wet and the cold,
> By the Grand Metropold, won't you give me a letter?
>
> Give 'im a letter
> 'Can't do no better
> Late Troop-Sergeant-Major an' – runs with
> a letter!'

With his poet's eye Kipling went to the heart of the matter. He saw the commissionaire as a long-engagement man whose inadequate pension following a life of colour and sacrifice had to be supplemented in the chill streets of Victorian London. Yet by no means all old soldiers were ex-Troop-Sergeant-Majors and by 1889, when Kipling wrote the poem, the Corps no longer required a large proportion of its men to stand in the wet and the cold. In 1887 no less than 863 of the 1160 men at Headquarters were in permanent employment in London. About 100 more were detached permanently in the provinces. There were only thirty-two acting messengers 'on post', like Kipling's Troop-Sergeant-Major, occupying busy locations such as the Royal Exchange, ready to take messages as required.[5] Admittedly, these were supplemented by some others at Headquarters who were also available for relatively low-paid temporary employment.

If less men were positioned on the streets than in earlier days, during the thirty years of the Corps' existence it had become a familiar institution to the vast majority of Londoners. The brisk, purposive movement of straight-backed men in their neat black uniforms criss-crossing the thoroughfares of London brought assurance and approbation. They were patently responsible and supported by a guarantee of good service from their parent organization. This guarantee helped to increase the demand for them which at exceptional times, like that of the Jubilee, they found quite impossible to satisfy. During these celebrations as much as £1 a day was paid, compared with the usual rate of £1 to £1 10s a week. In normal circumstances their rewards were understandably lower but they had now assumed a broad and strangely varied range of duties. More than 150 were doorkeepers or checktakers in the theatres or music halls, many more acted as messengers in banks and offices. There were commissionaire cashiers, drill instructors in schools, deliverers of circulars and private messages. Some acted as guides to strangers visiting London and selected staff-sergeants went as special messengers to the Continent on journeys of 1000 miles carrying valuable property and papers. Others were attendants in hospitals, day or night watchmen or temporary caretakers in private dwellings for which a man and wife earned no more than 2/- a night. On the other hand on 'settling day', when valuable bonds and securities were transferred from one stockbroking firm to another, members of the City detachment earned up to 10/- a day. A notable employer at this time was the General Post Office which employed about 120 men to deliver night telegrams; for this they earned 24/6d a week. The average weekly wage was 26/-, temporary employees earning less and the well-educated men considerably more.[6]

Quite apart from the regular sight of these uniformed men undertaking their many and varied commissions, the Corps added to the pageantry of the City in other ways. From 1862–1875 its Band performed daily in St James's Park, the only uniformed band to do so regularly in London. This was despite the sporadic objections from nearby residents. Edward Walter stonewalled these successfully, largely because he enjoyed, and carefully cultivated, the powerful support of the Gladstones who loved the music. In 1875, though, fresh objections from Carlton House Terrace led to a question being raised in the House of Commons and, despite a furious interchange of letters between Walter and the Home Office, the Band's activities were henceforth restricted to a few ceremonial occasions.[7] A notable one was at the Lord Mayor's Procession. Still forty strong, the band continued to be in great demand at private festivities and parties and held an instrumental concert in Headquarters from six to seven every Tuesday and Friday. This was open to friends and commissionaires' children.

On the second Sunday of every month Londoners could see the Corps and its Band on Church Parade, the muster taking place in Westminster Hall, while on the nearest Sunday to Waterloo Day there was a general muster of the Corps, normally in the grounds of the Royal Hospital, Chelsea, where the men and their wives and children were all reviewed.

London witnessed another ceremonial on the death of a commissionaire. In the case of sergeants and staff-sergeants the Band attended the coffin as far as the relevant London railway station, while a staff-sergeant was accompanied by another staff-sergeant, four sergeants, four corporals and twenty-four men to the interment within the Corps plot in the London Necropolis at Brookwood. When the Corps Sergeant-Major died no less than 100 men and all his non-commissioned officers accompanied the body to Brookwood. 111 Corps interments took place at Brookwood between 1859 and 1908 and many more commissionaires were accompanied by their uniformed colleagues to other London cemeteries.

The nub of regular Corps activity was, of course, its Headquarters; from here there occurred constant traffic of commissionaires not only from Headquarters Building itself but from premises all round it occupied by single men, while in two blocks of the new Sandringham Building there lived a colony of married men. At the top of the Headquarters was a large canteen/recreation room which was considered by many the best club in London. On more than one occasion Edward Walter referred to his Corps as a form of club and there was no denying the success of the canteen in the '80s and '90s. In fact in 1898 it became a subject of complaint from the Board of Inland Revenue for being open longer than the stipulated hours on Sunday mornings and being frequented not only by commissionaires but by civilians, including some disreputable ones. The Founder's response was predictable. In a furious salvo he said that in general he wished his men would 'keep themselves to themselves' but that 'a commissionaire has as much right to take his friend into his Club as I have into mine'.[8] Not for the first time he compelled the authorities to back down but the canteen keeper was dismissed and the system revised. Whatever the circumstances there is no doubt a very large number of commissionaires and other respectable citizens of London appreciated the hospitality of the mess over a number of years. After 1898 in a rather more restricted way they continued to do so.

Nothing could be guaranteed to rouse the wrath of the Founder more than a slur on the character of his Corps. Standing as it did in the eye of the public, nothing must bring into question the principles on which it stood – honest application, integrity, self-reliance, high standards of conduct and loyalty. Preserving and enhancing these required unremitting vigilance and continuity. This last factor concerned Edward Walter, particularly after his return from India. How could the Corps he had created, shaped and led for so many years continue in the path he had set? The thoroughgoing regulations helped to point the way but the obvious solution lay in the appointment of a successor whom he could imbue in its traditions and train to his own standards, a man whose thought processes matched his own.

To a Walter this would mean another Walter. The obvious successor was his son but he was making a successful career in the Army and was Commanding Officer of the 7th Hussars at the time of his father's death. Nothing must prevent him achieving command of his Regiment, something Edward Walter denied himself by retiring early, although the virtues of true regimental spirit inspired much of his work. The chosen successor was his nephew, Frederic Edward Walter, thirteen years older than his son and also a regular officer. He was persuaded to retire from the Royal Artillery after twenty-four years' service and to join the Corps, which he did at 43 years of age. Initially Frederic was made Officer Commanding the Leeds Division although within the year he was appointed to Headquarters as '3rd Officer'. His was to be a long apprenticeship stretching from 1891 to 1904, during which time Edward remained in full and sole control, although his absences due to ill health tended to grow longer. In 1895 Frederic was given Power of Attorney in the Founder's absence and in 1903 we find him reading the Commanding Officer's Report on behalf of his seriously ailing uncle, who died on 26 February, 1904.

By 1903 the long Victorian age had passed. (the Queen died in January, 1901) and the British Empire and its military establishment had been involved in a bitter war. The Boer War lasted three years, far longer than almost anyone had predicted (except perhaps Kitchener) but that is the nature of wars, particularly modern ones. It had cost over £200 million and 20,000 lives. To bring about a successful conclusion there had been a massive increase in the military establishment, including detachments from the Dominions. It revealed that in some respects British Imperialism was found wanting, as was the size and efficiency of its armed forces. During it were sown many causes of the terrible world conflict to come and a new sense of vulnerability. New security was required through a reshaped and realigned military establishment in the early 20th century. In the way of major conflicts over the preceding century it was to be expected that increased attention would be given (even temporarily) to the problem of the exceptionally large numbers of men returning from military to civilian life. It is for the next chapter to consider the nature of the initiatives undertaken to help returning servicemen and the part the Corps, under new leadership, was to play in the changed situation.

CHAPTER NINE
Aftermath of War: Under New Management

The Boer War marked a new peak of patriotic fervour and an extended interest in military affairs. This helped the merits of the British soldier to grow in public knowledge and estimation. It was essentially a land conflict and, as such, it caused the Army, in particular, to undergo rapid expansion with more civilians entering or rejoining its ranks than in any previous conflict. For the first time a fair proportion of their employers undertook to keep open the civilian posts vacated by reservists who returned to the colours.[1] Yet the enthusiastic but uneven scramble to the flag was far surpassed by the swiftness and concentration of the men's return. In 1902 no less than 64,445 soldiers were discharged with good characters and the total number coming home in that year reached 70,000.

The problems facing the employment agencies were multiplied by this formidable combination of very large numbers and short time scale. The National Association, for one, had made careful preparations for the human flood by attempting to extend its vacancy lists. It arranged the printing of 12,000 posters and circulars which appealed to the patriotic spirit of employers. Other societies like the Soldiers' and Sailors' Families' Association and the Soldiers' and Sailors' Help Society which were traditionally, and extensively, involved in welfare matters now turned to the serviceman's greater contemporary need, his successful employment on release.[2] In practice these generally tended to work in conjunction with the

National Association on employment finding. While much was achieved, many soldiers remained unemployed to suffer the inevitable hardships of this condition. In the year ending March, 1903, the National Association claimed to have placed 13,909 men, three times those of the previous year, but still only half the number of reservists registered with it.[3] The Inspector-General of Recruiting estimated that over 45,000 men found employment in that year. This was very probably an overestimate, and their average weekly wage of 19/- demonstrated the low level of such employment. Even taking the figure at face value it left over 24,000 men unemployed. Due to the needs of these unemployed servicemen, many of whom quickly became destitute and half-starved like their forebears in earlier centuries, fresh employment initiatives were adopted.

New initiatives apart, there were still the older voluntary organizations. How far, for instance, did the Corps of Commissionaires mirror the urgency felt by other agencies and to what extent could it help meet the challenge of the age? Plainly it had neither the flexibility nor the financial means (short of mortgaging its assets, a most dubious and unconstituitional process) to markedly increase its range of operations, let alone double or treble its normal level of activity as the National Association had done.[4] As a result, while undoubtedly most conscious of the war, the Corps' hierarchy tended to view it in relation to their mature organization with its well-established pattern of activity relating in the main to pensioners rather than reservists. They did not see it as something which presented them with opportunities for dramatic or markedly different approaches. During 1899 and 1900 the Commanding Officer offered accommodation at Headquarters to members of the Colonial Forces who were subsequently billeted there. By way of thanks they presented him with an illuminated address. In his response he took up his long-held theme of Corps growth overseas, emphasizing that 'if at any future time they desired to have a similar system established in their own portions of the Empire we should be happy to assist'.[5] To the end Edward Walter remained the visionary who looked beyond contemporary events to consistent long-term developments. However, no definite projects were discussed at this time and such comments were of little direct relevance to most of those soldiers returning home at the end of the war. During the war the Corps spent much of its time and energy in helping to place the wounded, but the war's most important effect on it was in the numbers of commissionaires who left to rejoin the colours. These totalled 353 men, some 12% of the roll. While this was a cause of much pride to Edward Walter and his officers it also caused various of the Corps Funds to suffer a considerable loss. In this respect the Corps was unlike the other employment bodies. While they experienced reduced employment activity during a war, it found itself directly diminished both in terms of the numbers of men available for civilian work and in respect of income from both the commissionaires and their employers. Fortunately, by 1901 some new recruits had been admitted and a third of the men had returned from overseas, while internal economies further minimized the damage. Expansion was able to be resumed according to the long-term pattern and a new out-division was opened in Bristol.

1902 was the year of maximum need for returning soldiers but the Corps neither possessed the elasticity of structure nor a large bank of potential posts to make a telling contribution here. This was demonstrated when the rapid return of commissionaires from active service resulted in those men who had stood in for them losing their jobs. One might have thought some prior arrangements could have been made to avoid such a humiliating and uncertain situation for those who had filled the gaps in the ranks caused by the war, and at least the Executive Committee showed itself fully aware of their position when it paid tribute to their admirable patience and good temper when kept waiting for new employment. However, first loyalty was unquestionably towards those men who went on active service and the problem was exacerbated by depressed industrial conditions which were a concurrent feature of the late Victorian and Edwardian years. From the vast outflow there was no shortage of good potential commissionaires, men much needed after the interruption to normal recruiting but 'it was a matter of deep concern to the officers that they were unable to admit (men)' because of the adverse industrial situation. In his Annual Report for 1902 the acting Commandant said he was confident that 'the difficulty which so many men experienced in obtaining work (in 1902) was not due to the lack of good will on the part of employers, but entirely to economic causes'. If accurate, such a thought offered little consolation to those hoping for admittance. The same Report expressed much pleasure in the maintained progress of the Corps. Undoubtedly it had been a year of recovery and consolidation; all had now returned from active service and, despite the depression, additional growth had occurred. The increase for the year, though, was limited to 76 men, just .001% of the total outflow. While the Executive Committee might refer to this as notable, 'considering the superior qualifications required by such a body of picked men', the field for selection could never have been larger. At this time it was the lack of new posts rather than any shortage of good men which inhibited expansion.

The fact was that the Corps did feel itself less affected by the war than the other agencies. This did not necessarily mean it was unfeeling. But its principal concern was to preserve its organization for serving commissionaires and to be capable of readmitting those who had temporarily left its ranks to fight, rather than make provision for the untold thousands of ex-soldiers flooding onto the labour market at the end of the war. In any case the majority of these were young men who would not normally be considered as potential commissionaires. The Corps had come to accept that expansion in the United Kingdom would be slow, without some massive and undreamed-of financial windfall and a substantial change of policy. It had come to accept the necessary rejection of many applicants either because they were below the high standards set or because market forces prevented it having many extra posts to offer. In this context the position during 1902 was not dissimilar to previous years, even if the numbers of aspiring entrants were much larger than normal.

One can only conjecture how far a vigorous Edward Walter might have taken advantage of a situation so likely to benefit both his Corps and the ex-serviceman on whose behalf he had worked for so many years. It would have required undertaking nothing less than a major new crusade on the Corps' behalf. Sadly, by the end of 1902 he was terminally ill. But he was still unwilling to relinquish the reins of power, even to another Walter who had fully earned his confidence and trust over a considerable period. This was a most unfortunate time for a grievously ailing leader and a caretaker chief executive. In the event 1903 was an unnoteworthy year for the Corps, the roll being increased by the relatively modest figure of ninety-eight men. On February 26, 1904, Edward Walter died, and the loss of the Corps' Founder and pilot for forty-five years was commemorated with considerable pomp and meticulous organization. While the preceding two years were most unlikely to have witnessed great change the tendency after his death was to adhere strictly to the traditional groundrules. On the Saturday following his death Frederic wrote a long letter to his officers and commissionaires, ending with the exhortation: 'He is gone; and it now remains for those to whom are committed the interests of the great institution which he founded, to see that the principles on which its organization is based are strictly observed.'[6]

Personally selected and brought into the Corps by the Founder, then drilled for twelve years in its régime and role, Frederic was hardly likely in his initial period to set such a pattern aside, if only out of respect for his uncle. This industrious, conscientious and sensitive man of middle years had other reasons not to set about changing the mould. Apart from his own conservatism and deep sense of tradition, he would never find himself in the dictatorial position of his uncle vis à vis the Executive Committee. Edward had believed that the role of Commanding Officer should be held by someone of independent means in order to retain his independence of action. Although he had understandably come to recompense himself for his Corps' work through a fund which he set up for the purpose in 1887, the Executive Committee never felt itself justified in discussing this question with him. Frederic was not to enjoy the same financial independence and would never exercise the same autarchy as the Founder. On his succession the Administrative Board voted him an official salary of £450 a year (including travelling expenses) to be increased from time to time as it felt fit. In 1905 the whole question of Corps control and the administration of the Officers' Endowment Fund was redrafted; this resulted in considerably more power for the Executive Committee, while in 1906 the accounts were audited by a firm of Chartered Accountants for the first time. Frederic could never defy the Executive Committee in the purchase of property as his uncle had done before him.[7]

The main outcome of this changed relationship was for the Corps to continue on the path so clearly ordained by the Founder. An Executive Committee of distinguished and busy men, meeting relatively infrequently, as it did at this time, was hardly likely to initiate sweeping changes to a successful institution. Ample proof of its success appeared to be manifest during the next few years.

On 16 June, 1907, the Commissionaires' annual parade was honoured for the first time by the presence of a reigning Monarch, Edward VII, who carried out the inspection. This followed four consecutive years of good recruitment. In 1904 the Corps passed the 3000 mark and increased by 120 men. The following year saw the greatest annual increase to date, 176 men, who were, in the Commanding Officer's words, 'of the best class'. During the next two years, this figure was itself easily exceeded and the ranks grew by over 200 annually. Labour troubles brought growth down during 1908 to below 100, but this year was marked by the establishment of a major new employment organization for ex- servicemen based closely upon the aims and practices of the Corps.

The new body was called the Veteran's Corps and, like other voluntary bodies, it came about through the initiatives of Army Officers. James Malcolm and Major Arthur Haggard were members of an Officers' Club in Hand Court, Holborn. They and other members, painfully aware of the great difficulties still experienced by many ex-servicemen in finding employment, determined to help them. The result was the establishment of a non-commercial company called The Veteran's Corps and Employment Bureau Ltd. This was guaranteed by a Council upon which, in the traditional way, were enlisted distinguished ex-officers, while day-to-day affairs were controlled by a Managing Committee. Affinities to the Corps of Commissionaires are not difficult to see: the Chief Executive Officer was called the Commandant and the Veteran's Corps was a uniformed disciplined body with a rank structure which insisted, like its older counterpart, on its members having good character records and being up to a very high standard of personal behaviour whilst in the ranks. Men were accepted for the same range of posts: as commissionaires, messengers, reception and enquiry clerks, gatekeepers, storekeepers, factory and dock police, time-keepers, bank and stockbrokers' messengers, resident caretakers, club stewards, porters, liftmen and watchmen. They were charged a set fee when obtaining full employment or 5% of their wages when on part-time duties, but employers were never charged. Like the Commissionaires' Corps each man acted under a Fidelity Guarantee and up to the First World War it had an insurance scheme to protect its members during illness or retirement. As with the older Corps it formed its own band of drums and pipes and held an annual parade, this time in the grounds of the Honourable Artillery Company. There was one main difference between them: the Veteran's Corps was generally run on far less rigid lines. From an early stage, if employers required it, men were provided in plain clothes; its rank structure was more flexible and it happily recruited reservists. This more liberal attitude towards reservists caused it to have fewer recruiting problems. An unavoidable brake on progress for both was the slow growth rate of any such hierarchical institution but within six years its roll stood at just below 1000. This was immeasurably faster growth than that achieved by the Corps during its early years. While its more relaxed approach could account for some of this, it also demonstrated the continuing and growing demand for non-skilled but highly trustworthy men from the two services, a requirement which both Corps were attempting to satisfy and which the Corps of Commissionaires had gone far to develop.

Having finally assumed control, one can understand the new Commanding Officer's justifiable satisfaction in his well-ordered and successful organization whose example was followed by another London-based body of ex-servicemen. His pride comes through in successive annual reports at this time and when walking through London past his own and 'rival' commissionaires in their smart uniforms one could forgive him for thinking that every ex-soldier who was good enough, and had a mind to, could become a commissionaire. In reality, despite the expansion of both bodies, they were still only catering for under 2% of the total outflow in 1910.

The years following a major war, especially when unexpected setbacks are encountered, are traditionally ones of reappraisal. The Boer War was no exception and certain of the military reforms had a direct effect on re-employment in civilian life. The reforms instituted by Richard Haldane who became Secretary of State for War in 1906 led to the formation of an expeditionary force to cope more adequately with future overseas challenges and the creation of a Territorial Army of Reserve to support the Regular Army. While unparalleled numbers of reservists had been used in the Boer War, inadequacies had too often appeared in their training, while the calibre of such soldiers had often left much to be desired. Haldane recognized the importance of good employment after colour service both for the success of his reserve Army and for the effective recruiting of the regular forces. To this end he appointed his Secretary to the War Office, Sir E.W.D. Ward, as head of a committee to examine the civilian employment of ex-soldiers and sailors.

The subject of ex-servicemen's employment had occupied a number of committees from the mid-19th century onwards but the 1906 Ward Committee was unquestionably a most important one. Its earlier counterpart was the 1877 Select Committee on the Civil Employment of Soldiers and Sailors. Edward Walter had been a leading witness on this committee both as the Founder of the Corps of Commissionaires and as the leading exponent for improved employment opportunities through the granting of minor vacancies in Government Departments.

In 1906 the position was much changed. Once more the Corps was called upon to give evidence, although the Commanding Officer, being indisposed, seemed content to delegate his role here to the Commanding Officer of the London Division. His evidence appeared to carry little weight compared with that of the other Agencies. In truth, the 1906 Committee saw the Corps as a major employer of servicemen and little else; certainly it was no longer a leading exponent in the struggle for more posts within Government Departments. Both the position and the thinking behind the serviceman's return had changed during the half century. The contemporary employment problem was too large for slow-growing organizations like the Corps or its carbon copy, the Veteran's Corps, especially if the former exercised a strong bias towards pensionable men. Not only were they generally too exclusive for the average reservist who was unwilling to subject himself to such rigid control, but, arguably, he now needed a wider range of opportunities, including posts of a supervisory nature, and, in an increasingly mechanical age, others which relied on a range of technical skills. Motor driving became a popular option here. The Ward Committee must have seen the fiercely independent Corps of Commissionaires, which maximised men's military virtues rather than their growing technical skills, as something of an aberration from the reservists' main need in the early 20th century. Yet the restricted implementation of the Committee's most important proposals ensured a continued need for all voluntary bodies, not least the uniformed ones.

The Committee raised anew the fundamental questions of the serviceman's return to civilian life: reserved occupations (no longer strongly advocated by the Corps but by the National

Association), the attributes of a soldier or sailor, how he could become better fitted for civilian life and the need for more thorough-going employment arrangements than the current 'benevolent operations of voluntary associations'.

Its main recommendation on employment-finding was for the creation of a Central Employment Organization. This did not materialize because the Treasury flatly refused to grant the necessary funds requested, £16,000 for the Army and £4,000 for the Navy 'to form the nucleus of the income of the Central Association'.[8] It is also true to say that the proposal met with opposition from the smaller job-finding agencies who were opposed to being absorbed in a greater (National) Central Association. With the failure to establish a Central Employment Association, the Haldane Army Reforms embodied a clause in the Territorial and Reserve Forces Act which charged the new local Territorial Associations, among other things, with responsibility for the care of reservists and discharged soldiers. The Territorial Associations only began to assume these responsibilities in the years immediately before the First World War and so there remained no shortage of potential recruits for the two commissionaire Corps.

On the question of reserved occupations the Committee recommended that various public departments should be directed to fill suitable vacancies with ex-soldiers and sailors of good character rather than civilians, notably for clerks and messengerships and a large proportion of future vacancies in H.M. Customs. County and Borough Councils should be urged to reserve a number of their appointments, and the Police Forces both to give preference and raise their age limits for recruiting servicemen. Once more the results were disappointing. Although some Customs vacancies were made available, posts in other Government Departments were not and the Home Office refused to grant an age concession for the Police Forces. With respect to increased skills for civilian life, the Committee recommended that all soldiers should be required to learn some kind of technical work during their military careers which would qualify them for certificates of proficiency.

Arrangements should be made at Technical Institutes for the tuition of men in various trades and crafts, particularly towards the end of their service. Regrettably it was not until 1912–13 that three Central Schools for Military Technical Instruction were established at Aldershot, Portsmouth and Woolwich.

While the military establishment was considering the Ward Committee's proposals to improve servicemen's prospects on release, political developments were occurring in the civilian employment field which were also of significance to them. The first decade of the 20th century had been marked by a continued and relatively high level of domestic unemployment as the UK lost its traditional markets to its competitors, notably Germany and the US. Lloyd George attempted to cut at the roots of the working man's poverty by instituting reforms, most of which Edward Walter had long anticipated. The Old Age Pension Scheme (1908) and the National Insurance Act (1911) attempted to achieve for the nation what Walter's scheme had provided for his commissionaires when sick and in retirement – admittedly not when unemployed. Lloyd George's Trade Boards which attempted to gain reasonable levels of minimum wages were also in accordance with Walter's aim for all commissionaires in permanent employment.

Of direct relevance to all ex-servicemen were the National Labour Exchanges (1908) where civilians out of work and employers looking for men could register their requirements and hope to get in touch with each other. With the breakdown of the Ward Committee's proposals for a Central Employment Association of their own, many ex-servicemen, particularly ex-soldiers, continued to fare badly in their search for civilian employment. They needed all the help they could get and hoped to realize their hopes by registering with the National Labour Exchanges. In practice many were disappointed because the Exchanges were not primarily designed for the special requirements of men who had been away from civilian life for up to seven years. The National Association for one certainly viewed them initially as contenders in the soldier's employment field and of less benefit to him than the voluntary bodies devoted to his welfare and no other. 'The ex-soldier is already a loser by the competition of the new Labour Exchanges . . . . Work which used to be offered to the Association by Borough Councils . . . is often now given to the civilian Labour Bureaux.'[9] It therefore worked determinedly to co-operate with the Exchanges (as far as possible on its own terms) and to have transferred to it certain responsibilities for servicemen, for instance duties relating to the registering and recommendation of ex-soldiers for partial employment.

Sadly the outcome of both these military and civilian measures did not greatly improve matters for the ex-serviceman. The Ward Committee's recommendations, sound and thoroughgoing in themselves, were either blocked or seriously delayed while the civilian system of benefits for the unemployed, including Labour Exchanges, were not of prime advantage to them. The result was that ex-servicemen, and ex-soldiers in particular, would not be the immediate subject of other urgently-needed reforms and would generally experience more serious employment problems on their return to civilian life than average civilian working men.

The 1909 Report of the Poor Law Commission confirmed the continuing problems and the good fortune of some who were accepted as commissionaires: 'It is the men who have left the permanent situation afforded by the Army . . . who furnish the largest contingent of the floating population of the casual wards. What they aim at is to get permanent situations "of trust" . . . or as members of the Corps of Commissionaires. But only a minority of them secure such places. The others mostly sink gradually to the "casual" class.' Out of 1512 men who passed through the casual ward of a rural workhouse in the West of England in 1905, no fewer than 333 convinced the Master that they had served in the Army.

In a hard-hitting book called *The Ex-soldier by Himself*, published in 1916, the onetime gunner and ex-soldier's apologist, W.G. Clifford, described the sad plight of the reservist prior to the First World War in vivid terms. It was yet another proposal of the 1906 Ward Committee which had not been taken up that led him to publish his own *Employment Gazette*, full of pratical hints for job-finding. He was a stern critic of the War Office's inaction in respect of the ex-soldier. Clifford also commented on the various voluntary employment bodies and his remarks on the Corps of Commissionaires are illuminating as coming from a contemporary and informed observer:

'The Corps of Commissionaires concerns itself mainly with providing situations for pensioners . . . . This fine Corps comprises a unique body of disciplined men, it is old-established, has considerable funds and a deserved reputation among employers which places it on a pinnacle in a state of splendid isolation . . . . It supplies guaranteed and uniformed men of unimpeachable honesty and good character to employers . . . which make no great call on technical qualifications. . . . The Corps cannot do more than touch the fringe of the ex-soldier problem. This is so true that not one man in a thousand of the vast numbers now confronting us will stand a chance of becoming a commissionaire.'[10]

It is interesting to consider how far Clifford's remarks were a fair commentary on the position of the Corps shortly before the First World War. Progress between 1909–13 had been impressive, the Corps increasing its roll by 1,000 men to 4,750, but it still primarily catered for the long-service men rather than the reservist majority. Both its exclusiveness and priorities of recruitment were indeed illustrated by the Commanding Officer's reference in 1908 to the newly-formed Labour Exchanges. He called them a much-needed system which would not affect the Corps, 'but should prove very beneficial to Reservists and would doubtless afford many advantages to the great majority of soldiers'.[11] With regard to Clifford's point about its considerable funds, a valuation of Corps property in 1908 found it to be far in excess of the cost price, but prospects for expansion relied primarily on the Endowment Fund and this grew slowly. The pinnacle of excellence for which it stood was referred to by the Commanding Officer in 1909: 'I believe there does not exist in any other country an organized body of veterans . . . at all comparable to this.' As a most conservative Walter he was not afraid of its relative isolation, referring to the value of commissionaires 'whose minds and manners are not disordered by the spirit of the age'. Unquestionably its permanent members just before the First World War were of a high calibre and happy with their lot in the Corps; just 1½% were dismissed in 1912 for poor conduct. In 1913 95% were with specific employers and no longer in the streets waiting for commissions.

Clifford's central observation and most telling criticism was that the Corps did no more than touch the fringe of the soldier employment problem. He saw it as incapable of change and it is unquestionably true that Frederic Walter concentrated on continuity as such – continued excellence on the pattern set by the Founder. The organization itself had understandably taken precedence over the wider question of servicemen's employment problems as a whole, especially those of reservists. One could hardly blame it, however, for the spectacle it gave of healthy responsible ex-servicemen, an impression which helped to cloud in the public's mind the reality of the many unfortunates occupying the casual wards of county institutions. Equally important, the Corps continued to grow – as did its imitator – along its old lines, and did not, as many ex-servicemen wanted, concern itself with the pursuit of occupations demanding technical skills. The basic question which Clifford raised was whether the Corps would adapt to a new age, preserving its standards but providing for the new needs of the 20th century, as it had done so well for those of the mid-Victorian times in which it had been established. Could not Edward Walter's unyielding standards of honesty, service and loyalty be applied to the expanding security and service roles of the contemporary age? A war of undreamed-of proportions over four long years was to bring the population as a whole into the armed forces; it was to threaten the very foundation of the nation – and the Corps itself – and to cause great change in the nature and pattern of the civilian workforce in Great Britain.

CHAPTER TEN

Lost Opportunities in a Time of Unemployment

On the outbreak of the great war 'to end all wars' no less than 2145 men, virtually half its roll, left the Corps to join the two services. Although they were generally men of mature years they included all the younger and fitter members. This caused problems for the Corps in being. The exodus left behind a less effective body of men, a fact demonstrated by an exceptionally heavy sick list during 1914 and a number of deaths far above the average. During four long years normal recruitment simply did not operate. New commissionaires were either men above military age or men who had been wounded or physically downgraded as a result of the conflict. Of 277 admissions during 1915, 127 were for wounded men (the downgraded were not identifiable); in 1916 there were 184 wounded out of 273 and in 1917 the wounded totalled over 60% of the intake. During these years the Corps simply could not undertake tasks requiring robust men, although stories of gallant attempts by the wounded to do so were legion. One legendary double amputee, having lost an arm and most of one leg, apparently returned to duty in the first instance complete with crutch and empty sleeve. At the select fashion house where he took up his post he received unparalleled courtesies from the public and experienced no problems in carrying out his role. Other commissionaires, though, had less obvious disabilities which nonetheless seriously affected their physical well-being.

The exodus caused another serious problem, and one which

had been experienced during the previous war, namely a financial crisis. The Corps no longer enjoyed the subscriptions of those who were absent, except in a few cases where former employers generously continued to contribute. Due to the seriousness of the financial implications, almost half its regular income being lost, the Executive Committee decided it was right and proper to approach one or more of the great funds which had been raised or supplemented by the Public for the benefit and welfare of soldiers and sailors. The result was bitterly disappointing. The Secretary of the National Relief Fund disposed of the Corps' request in very few words, being 'directed to inform you that the Committee . . . regret that they are unable to accede to your application for a grant in aid of funds of the Corps of Commissionaires'.[1] The plain fact was that there were infinitely more urgent calls on such funds than for a not-impecunious organization responsible for the employment arrangements of less than 2500 old soldiers, even if one added a further 2000 or so once the war ended. The justice of the decision became more evident when the Commanding Officer proudly announced the completion in 1916 of a lofty and capacious building erected by the Trustees of the Building Fund for occupation by the New Zealand High Commission. In the same year the Executive Committee was also able to report that new measures of belt-tightening, of internal economy necessitated by the wartime conditions, were working smoothly. These included temporarily leaving many out-divisions under the command of Sergeant-Majors rather than Officers.

Such decisions enabled the Corps to function satisfactorily prior to the resumption of normal activities. Yet the titanic military struggle which then overshadowed everything else was to influence the wider post-war employment scene. The urgency and scale of the war's demands accelerated the pace of mechanization; the virtually unrestricted expenditure of the nation's capital was to be followed by inevitable retrenchment and slackness of employment in the early post-war years. Even the labour force itself had been transformed by the entry of women to an extent hitherto undreamed of. All of these things would have their effect on the continuing capacity of the Corps to provide employers with unskilled but carefully selected men. External factors apart, over these years the Corps had certain serious internal problems of its own.

In particular it experienced a leadership crisis. At the beginning of the war the Commanding Officer expressed an understandable wish to retire at the end of 1917 just before his 70th birthday. During 1916 the President of the Executive Committee lamented that, 'the time is approaching when Major Walter feels he should hand over the reins to a younger man'. It was hardly a surprising decision. The succession had been carefully prepared and the Chairman went on to say, 'I hope he may be succeeded by Colonel Richard Walter whom you all know and that we may thus be able to congratulate the Corps on its command remaining in the Founder's family.' Sadly this reference to Colonel Richard Walter had to be deleted from the annual circular which normally carried in full the Commanding Officers' and Presidents' Reports. The Founder's son was never to assume command. The Committee minutes reveal that he attended the Annual General Meeting of the Governors on 16 June, 1917, when he was offered the post. His response was to ask for time to consider, as he was engaged in the Ministry of Labour and he also expressed some reservations on the very complicated funding arrangements of the Corps. While he still continued to attend Governors' Meetings, he was plainly not anxious to become the Corps' Commanding Officer at this time. The disappointment of Frederic, who had just cause to expect a well-earned retirement after long years of service, can well be imagined. The prevailing uncertainty about the future leadership was illustrated by a letter sent from the Corps' solicitor to the Treasury Solicitor. This asked whether the Corps could be taken over by a Government Department as an Incorporated Society. The answer from the Treasury was that 'it would not appear to be within the functions of any Government Department to take over the administration of the Corps and that in any case the advantages of its organization on the basis of a private enterprise appear to be such as to make it undesirable to make a change in this respect.'[2]

The Corps letter to the Treasury was sent in October, 1918, a time when it needed strong and clear leadership to help cope with the vast numbers of servicemen who could be expected to return shortly. From 1918 onwards there was considerable general discussion on amalgamating welfare and employment facilities for ex-servicemen. Earl Haig's British Legion partially succeeded in this object, particularly with regard to welfare arrangements. At no time, though, was the Corps of Commissionaires considered a potential candidate for such unified ventures. Its initiative to the Treasury was the direct result of an internal crisis arising from difficulties about its future leadership.

In any case the difficulties of merging such a unique and hitherto fiercely independent body into a conglomerate would have been formidable. Further evidence of the Commanding Officer's dilemma as to the future of his post came in a suggestion he made in December of the same year (before the official reply from the Treasury Solicitor had been received). He explained that it was probable the Governors would want to retain control of the Corps in their own hands and suggested that the office of Commanding Officer should be administered by a Committee of five officers, including a representative of the London County and Westminster Bank who would be concerned with financial matters only. One hesitates to imagine what the reaction of the Founder would have been to this naive proposal of governance by committee but it illustrated vividly the diminution which had taken place in the position of the Commanding Officer in relation to the Executive Committee. Although the minutes reported that the suggestion was very favourably received by the Committee they wisely did not proceed with it. Frederic Walter was to continue at the helm for another twelve years, making forty years' service in all, the first thirteen as Secretary and understudy to the Founder, the next twenty-seven as Commanding Officer. Prospective candidates were brought in to succeed him but problems continued to arise. His eldest son, Frederick Henry, on retirement from the Royal Navy took up the traditional heir presumptive's post of Corps Secretary. However, he was forced to quit his duties in 1926 on grounds of persistent ill- health. Frederic's third son, Edmund, became Corps Secretary from 1927 onwards and acted as Deputy when his father was away sick, but he only assumed command after his father's sudden death while still in harness.

The leadership problems which occurred from 1917 onwards inevitably had their effect on the Corps. In its first seventy-two years it was commanded by just two men, the second of whom attempted unsuccessfully to shed his duties when approaching his seventieth year, thirteen years before his eventual death. Despite his unceasingly high sense of dedication and conscientiousness through these thirteen years, years when most men would have retired anyway, he was a reluctant leader. In any case Edward and Frederic Walter were very different. Edward thirsted to establish his personal reputation through helping ex-servicemen at large. Frederic was far less ambitious and enterprising and in any case, prior to his accession, the Corps had yielded its premier role to the National Association and settled into a limited growth pattern. He was by inclinaton an intellectual rather than a man of action, contributing elegant prose and verse to *The Times* and delighting most in giving erudite and striking speeches at the annual inspections of the Corps. These speeches revealed his bewilderment and deep suspicion of the current age. In 1922 he called it an age of 'wonders and blunders'; two years later he cited Hamlet observing, 'There is something in this (age) more than natural'. About one thing he was certain, far more certain than his uncle, who admittedly had worked with more difficult material, namely about the sterling qualities and worth of the ex-soldier and sailor. Individual commissionaires loved him for this faith. Yet a number of things – his natural inclinations, coupled with advancing years and a long wish to retire, his antipathy towards the rapid and often contradictory developments of the inter-war years – did not equip him to consider any realignment of the Corps to meet contemporary challenges.

From the time of the Founder's death until the Great War little innovation had occurred, although there had been some geographical expansion in the out-divisions. During the twenties there was virtually none at all. With the accession of Edmund Walter certain much-needed, if relatively minor, reforms were undertaken. Yet he too continued to follow, almost slavishly, all the original precepts of the Founder, some of which were devised for an age which had passed. In 1937, shortly before the Second World War, when the Corps was labouring somewhat under the image of its older men, complete with their highly waxed moustaches, he commented on the fact that more short-service men had joined during the past year. Yet he was quick to explain it away as 'possibly only a temporary phase' – not something to be encouraged or approved of. If, prior to the First World War, W.G. Clifford had seen the Corps as a rather magnificent relic of an earlier period, how much more apposite was this judgement about a still unchanged organization just prior to the Second?

That the Corps maintained and even extended its strength owed much to external factors, notably the very inconsistencies and turbulence of the inter-war years which Frederic hated so much. The political and economic breakdowns following the Great War brought unemployment to the Western nations as a whole. Not until the later thirties, with the beginning of re-armament for another war, did industrial activity start to accelerate in this and many other countries.

In such circumstances, where skilled and unskilled alike were unemployed, the uniformed bodies who offered employers ex-servicemen of high calibre found themselves with a most important role still to perform. In 1924 a demonstration of continued faith in the demand for such men to perform security and carrying duties came in the formation of yet another body for this purpose. The Commissionaires' and Veteran's Corps found themselves joined by the Royal Artillery Association's Uniformed Staff Corps which primarily operated in

London but also had out-stations in other cities, particularly Manchester, Liverpool and Sheffield. Like the others, it recruited men of good character but did not have the stringent regulations of the oldest Corps and it made no regular charge (save for uniform upkeep) from either its members or their employers. From the outset it grew steadily.

The popularity of such traditional bodies could also be partly explained because the savage cuts in public spending, during a prolonged period of depression, hampered any new arrangements for improving servicemen's civilian job prospects in other areas. The typical old sailor and soldier (joined now by the airman) found himself virtually no better off than in the earlier years of the century, with the attention of his erstwhile champion, Kipling, more concerned with castigating those whom he considered responsible for the war casualties, including his own son, than with the difficulties experienced by normal peacetime soldiers. In any case the returning old soldier was just another element in the vast army of the unemployed and an unwelcome one to many of the working population at large. During this period less discussion occurred about special arrangements for ex-servicemen than at any time since the founding of the Corps. The sensible and sweeping proposals of the 1906 Ward Committee and the later work of the two Committees which sat just prior to the war were lost in the general problems of unemployment.

Just one group of ex-soldiers was better off than before, namely the wounded and disabled. The huge casualty figures and their relatively high survival rates made special arrangements imperative here. Yet another objective of Edward Walter, taken up by him fully sixty years before and followed by Lord Roberts' Workshop Scheme, was to become reality. From 1919 temporary posts in Government Departments were to be reserved for disabled men and, through the King's National Roll Scheme, Government contracts were restricted to firms offering a percentage of their posts to such men.[3]

For the normal outflow few advances occurred. In the Army, Vocational Advice prior to release was the responsibility from

1920 onwards of the fledgling Army Educational Corps, but its strength was to be halved by the 1926 Geddes Axe.

Training for civilian life was provided by two special Army vocational centres at Hounslow and Catterick, but their facilities were limited and attempts to extend them were refused on financial grounds. At Catterick agricultural courses were offered to prepare servicemen for emigration. Under this scheme parties of soldiers and their families sailed as emigrants to Western Australia until 1931 when the arrangements were halted due to financial cutbacks. In Aldershot there was a Building Trades Centre which offered courses of six months' duration but only a relatively few lucky men were eligible for selection.

In the employment field particularly, progress was bitterly disappointing. A Committee sitting under the Minister of Labour (Sir David Shackleton) recommended the establishment of a Central Committee comprising employers, the three Service Departments and the Ministry of Labour to help time-expired ex- servicemen. Its proposals were never adopted due to resistance from the Trade Unions and their antipathy caused other similar schemes to collapse. For instance, in 1921 a Cabinet Committee proposing co-operation between Government Departments for the employment of ex-servicemen throughout the building industry was abandoned because of such opposition.[4]

The overlapping and wasteful system of employment-finding for the Army by a combination of voluntary organizations and the County Territorial Associations continued. The idea of a Central Employment Association seemed further away than ever. The nearest one came in the inter-war years to centralization was in 1931 when a War Office Committee recommended the co-ordination of the voluntary job-finding organizations through the Ministry of Labour.[5] The voluntary bodies interpreted this more as domination than co-ordination and not necessarily as being in the servicemen's best interests. All soldiers were now advised to register at Employment Exchanges and, if they wished, with the voluntary associations too. By this time it was the National Association which had come more and more to represent the voluntary bodies as a whole. In 1920, for instance, it absorbed the organization which marked Edward Walter's entry into the employment field, the Army and Navy Pensioners' Employment Society. In 1922 it extended its facilities to all three services, thereby increasing its annual income and allowing it the opportunity to employ paid job-finders of its own. While the Association officially welcomed co-ordination with the Exchanges, it was in no doubt that servicemen needed voluntary organizations too, ones which worked expressly on their behalf rather than relying solely on employment services designed for civilians.

It was no wonder, then, that in such an adverse climate servicemen found the going hard. Great hopes were generated by a Treasury Circular in 1930 which referred to certain types of Government Employment being made available. In the event they proved to be low-level posts and the numbers offered were much lower than expected.

In some years the National Association could proudly announce that up to two-thirds of those who registered with it found jobs. However, of those who registered, the Naval and Air Force men invariably had less difficulty than soldiers and these employment figures included a considerable proportion of casual and low paid work. For much of the time only about 50% of those ex-soldiers who actually registered with it found work of some sort.

Under such difficult circumstances one would have been very surprised if the Corps had lacked recruits. Servicemen needed a job, almost any job, to save them from destitution. Many had greater technical expertise than their forebears, were better educated, had more self-discipline and were better briefed on the ways to seek for civilian jobs. Due to the adverse civilian job market which had operated over many years they were often men of the highest calibre. In their prime need for employment they flocked to all those agencies which might help, including of course the Corps and other uniformed bodies. When interviewed by the author and asked why they joined the Corps, most old commissionaires gave the same simple reply, 'Because we needed work'. Many were to give notable service to those leading civilian firms who recruited through the Corps and to stay with them the whole of their working lives. In Scotland their employers numbered, among others, such doyens as the House of Fraser, Maclay, Murray and Spens, George Outram, *The Scotsman*, Dewar's Whisky, White Horse Whisky, and The Royal Bank of Scotland. The Corps might have rather less ambition, less firm leadership than before but its continuing high reputation and the adverse employment climate would bring it no lack of would-be recruits.

The sad thing was how little it seemed able to take advantage of the situation. During the period the roll remained remarkably steady, whereas it could have been doubled or trebled if vacancies had been more energetically sought. On demobilization ten times the numbers of men who were admitted applied for membership, while in 1922 many hundreds of the best stamp applied, 'but the scarcity of situations prevented most from being accepted'.[6] From a total of 4694 in 1914, the roll declined to 4471 during the economic blizzard of 1922, rising to just over 5,100 during the upturn in the period immediately before the Second World War.

This latter total represented the Corps' high point in numerical terms, although for most of the inter-war years the totals were below the 1918 level. With hindsight the missed opportunities are not hard to detect. In 1919 the Commanding Officer regretted that in some districts a good many more men might have been employed on temporary engagements if uniforms could have been provided. One is forced to wonder, at this time of dire need for so many ex- servicemen, whether the rules might not have been amended, whether uniforms could have been borrowed or just hats, belts and armbands issued as a temporary measure. The period called for special efforts to publicize the Corps' work and, if possible, to extend its areas of operation in both functional and geographical terms. In fact publicity remained relatively muted until the early '30s; the

Corps' functions were virtually unchanged and priority was given to improving the property holdings rather than to geographical expansion. The mood of the leadership was hardly urgent: it regretted the would-be recruits who had been disappointed but expressed satisfaction in the performance of the Corps as a whole. The Chairman of the Executive Committee in 1921 reported that, 'The present condition of the Corps appears to me more satisfactory than we might have expected'. In the following year the Mess premises at Headquarters were extended, while the numbers were falling both in London and in the out- stations. Simultaneously some economies were being imposed on the out-divisions. In fact this potential area for growth had been the subject of rationalization and some retrenchment since the end of the war. In 1919 Nottingham was placed under the control of Birmingham and Edinburgh under Glasgow. In 1923 Belfast lost its Commanding Officer who was replaced by its Sergeant-Major. With energetic and dedicated Commanding Officers of their own, the out-divisions would surely have been better able to withstand the pressures of the early '20s and to take fuller advantage of the upturn in the later '30s. Parts of the country remained untapped. The long-intended Division for Wales had yet to be formed (it remained inadequately covered from Bristol) and East Anglia had no Corps representation. There was scope, too, for initatives in such places as York or Canterbury, the latter possibly in conjunction with the South Coast.

Even in the heart of the depression fresh opportunities became available. Because of prolonged hardships there occurred a lowering of public morality, particularly with respect to property. New organizations appeared in the security field. In 1935 a group of ex-officers bought out a tiny company of night watchmen in order to organize effective supervision of their town houses while they were away. They staffed it largely with reservists from Cavalry Regiments and called it Securicor. (The military connotation can be seen quite clearly in the name). Both Night Watch and Securicor used bicycles, but in

The Sydney Division of the Corps of Commissionaires, 1889: Edward Walter's first attempt to establish the Corps overseas.

A group of Royal Marine Commissionaires, 1896.

Major Frederic Walter, nephew of and successor to the Founder.

The Annual Parade of the Corps of Commissionaires at the Royal Hospital, Chelsea, 1889.

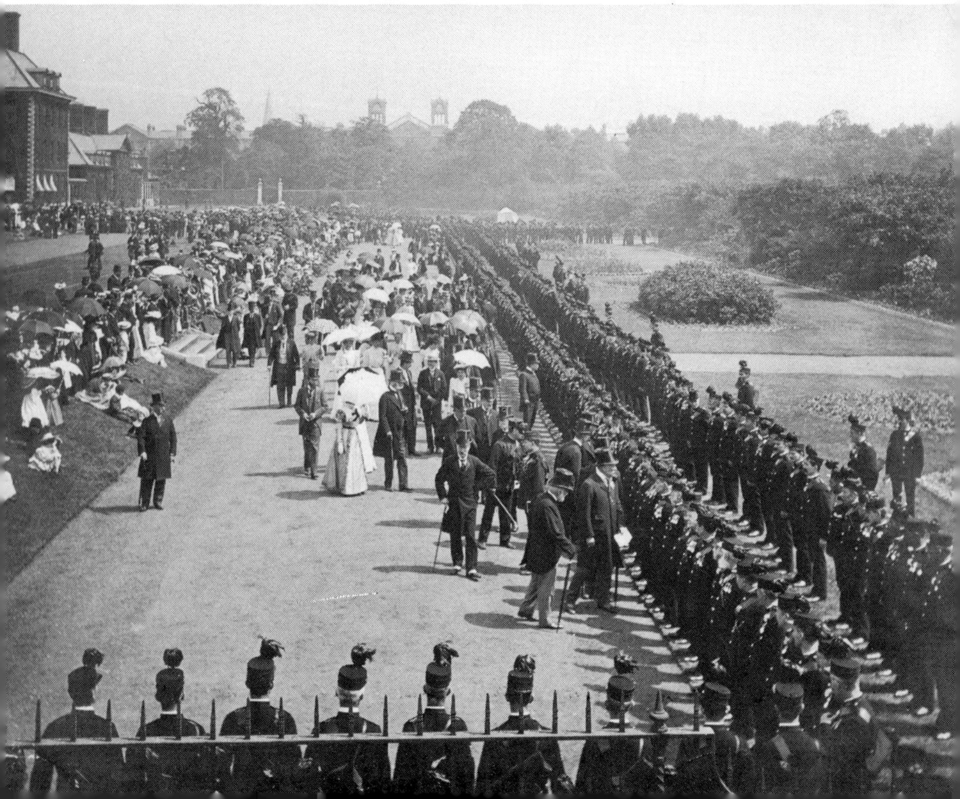

the post-Second-World-War era Securicor graduated to motor vehicles and started on the rapid path of expansion which brought it to a turnover of £236 million in 1983. While an organization such as the Corps would never have been able to take the risks open to a purely commercial company, it is puzzling why no initiatives were taken to use more modern methods to help perform its role. In the 1860s, when the Corps had virtually no funds, Edward Walter instituted a system of taxis. This was called 'Commissionaire Cabs', the full details of which were given in the Quarterly circulars of the time. A twenty-four-hour service from Headquarters was offered. It was discontinued after a relatively short time because it did not prove a commercial success, yet the enterprise shown was undeniable.[7]

In 1930 the Corps General Reserve Fund stood at over £17,000 and had been most healthy for some years before this. Such capital offered scope for expansion either in the extension of the out-divisions or even through some form of mechanization. It is true that neither the Veteran's Corps nor the Royal Artillery Staff adopted novel methods, but in their case the opportunity was hardly there owing to lack of capital. In any case they succeeded in expanding far more quickly than the Commissionaires' Corps, the Veterans reaching virtually half the latter's strength by 1938/9 and the Royal Artillery growing to over 1500 by the same time.

The idea of using transport, even bicycles, in the course of commissionaires' duties seems never to have crossed the minds of either of the two Commandants or their officers between the wars. Perhaps it was thought an undignified practice for commissionaires to ride bicycles. Edward Walter had certainly vetoed them pushing hand carts for this reason. Yet bicycles with box panniers in Corps colours or vehicles sporting the Corps badge would hardly have aroused such criticism. Unquestionably a great deal of time and energy was spent on the impressive ritual and panoply of Corps life, including the rebuilding and extensive refurbishing of Corps property, particularly the Headquarters Building. There were, too, the

Annual Inspections conducted by only the most distinguished. (Of the eighteen parades held between the wars, six were taken by members of the Royal Family, including King George V, and on seven other occasions the Inspecting Officers were Field-Marshals).

Neither Commandant had experience of, nor great enthusiasm for, commercial affairs, and mechanization would have required quite major changes. In 1931, for instance, 90% of the men, certainly the best men, were in permanent employment with civilian organizations. It was to these they owed their primary allegiance, especially firms who granted pensions after long service. Many of these commissionaires became 'company men' who only knew the Corps as a shadowy organization through which they had gained their employment opportunity, and for which they paid fully through annual subscriptions. Some might visit it just once a year to renew their uniforms, or less frequently if clothed by their employers. On certain occasions they might take part in the Annual Parade.

The other 10% were mostly over sixty years of age. (Some were younger men who supplemented their regular employment.) 'Our Sergeant' played a most important role in firms and organizations up and down the country, but it was 'Our Great Uncle Sergeant' who was chiefly in the public gaze. These took on the temporary public engagements performed by the Corps at exhibitions, soccer matches and race meetings. The latter provided a particularly important source of temporary employment and one probably much to the satisfaction of the Corps hierarchy. These elderly men undoubtedly performed with great credit; in their sparkling uniforms they represented authority; their politeness and sense of humour brought them willing co-operation, their grey hairs, medal ribbons and wealth of experience diffused possible confrontations.

Yet this group was fully tasked and neither sufficiently large nor physically robust enough to perform the higher-risk tasks of transporting valuable consignments at short notice around the clock. To undertake this role it would have required men specially recruited for the purpose and for the Corps to establish extended storage and maintenance facilities. It called for the reversal of a trend, recreating a pool of vigorous men at Headquarters constantly on call for such duties. Although the barracks remained full, by this time their facilities were antiquated and they were occupied by elderly, generally less energetic, bachelors whose rent was set at some way below the true economic level. Discounting the possibility of transport altogether, a pool of younger more vigorous men was needed if the Corps was to consider undertaking more sensitive roles, like guarding embassies or even undertaking the routine security tasks of some military installations. Between the wars no effort was ever made to build up such a contingent.

After 1931, when the 50-year-old Edmund Walter assumed command, the tempo of executive decisions quickened, but they were neither radical nor novel. They were concerned with two fundamental issues – the internal cohesion and efficiency of the Corps and maximizing its good external image. Virtually his first act was to set about bringing the Administrative Board of Governors to full strength. At a lower level he instituted a Corps Badge, in silver, for men with twenty years' exemplary service, a badge which they could retain on retirement. He encouraged the sense of Corps identity and Divisional loyalty through reunion dinners. In 1936 these were held in London, Manchester, Liverpool, Leeds, Newcastle, Bristol and Cardiff. An effort to reduce the 'Grandad' image came in his directive that Corps members should normally retire at 65, but this still left latitude for special cases, of which there were apparently many. Without quite ruthless weeding, the 'senior' image was to prove most difficult to change. In any case it was deep in the public's consciousness of a commissionaire. *The Times* exemplified this fact, for instance, during its report on the Annual Parade which was held in 1932. It wrote: 'The sun shone on the steady ranks of these men, few of whom were under 50 years old and many of them nearer 65.'[8]

Like his predecessor, Edmund Walter believed in making good use of publicity. He doubled the fidelity guarantee for commissionaires from £25 to £50 and this was favourably reported in the Press. In 1933 he appointed an officer (Lieutenant-Colonel R.C. Byrne) for the sole purpose of propaganda work. In five months Colonel Byrne visited 500 firms and over 2000 more in London during the rest of the year. In 1934 a projected film called 'Commissionaire' was given every assistance by the Corps and in the same year an anniversary Souvenir Album was published along with a pamphlet, 'What a commissionaire can do'. The appointment of Colonel Byrne was an important and long overdue step. The other initiatives, sensible and important as they were, tended to enhance the historic traditions of the organization rather than exploring new areas of operation.

The same conscientiousness and lack of imagination in the Corps leadership between the wars could be seen in the attitude adopted towards overseas developments on behalf of First World War veterans. Contrary to the Founder's enthusiasm here, the hierarchy showed little interest in Colonial Corps of Commissionaires and, if not downright suspicious, they were exceedingly cautious about any possible commitment the parent Corps might make towards them.

In 1934 a letter was received from a Corps of Commissionaires recently established in New Zealand. It was thirty strong in Wellington and twenty-seven strong in Auckland, but it had great ambitions and declared there were hundreds more willing to join if the establishment could be developed. The communications ended with the hope that it would be 'a great honour and privilege if some sort of liaison could be established between the present body at home and the New Zealand Division.' No reference was made to it in the Commanding Officer's Annual Report and there is no evidence of any positive response.

Three years later an initiative came from Canada, through Sir Charles Ratcliffe, relating to a (so-called) Corps of Commissionaires in Vancouver Island, British Columbia. The response from the Corps' Executive Committee was quite clear. They were unanimous in upholding the view that, 'under no circumstances can the Corps consider affiliation in any official connection whatsoever with any other organization working on parallel lines either in the colonies or elsewhere'. Their fear was for the independence of the parent Corps, with affiliation leading to possible legal claims upon its assets. Their care in preserving the position of the parent Corps was understandable but the nature of the response evinced a negative, indeed almost hostile attitude, to a *bona fide* organization which had already reached considerable proportions and which was eventually to become very much larger than the parent Corps. In 1937 the Canadian Government granted letters patent to the Canadian Corps of Commissionaires (it had been in existence since 1912) for it to organize Divisions in every Canadian city with a population of 40,000 or over.

Immediately prior to the Second World War, Headquarters received a communication from Australia, the scene of Edward Walter's first and only overseas initiative, revealing the formation of a Corps of Commissionaires in New South Wales, once more at Sydney. There were requests from other members of the Commonwealth for assistance in setting up their local Corps of Commissionaires, among them from South Africa and Egypt. One cannot help feeling that with active encouragement and assistance from the parent Corps more of these would have flourished and the Corps could have achieved true international, rather than primarily national prestige. After the Second World War dramatic growth did occur overseas but this was almost completely divorced from the founder organization.

The pattern of Corps activity between the World Wars remained consistent. Under careful leadership it kept strictly to the main course and high standards set by the Founder: eschewing radical initiatives towards growth, not eager to extend its functions in the security field, slow to change practices arguably no longer relevant in a changing world and apprehensive of the risk potential in the establishment of overseas Corps. During a period of continued employment difficulties, the leadership apparently considered any questioning of the Corps role inappropriate; it continued as a considerable and successful employer of ex-servicemen who in the main had reached pensionable age. At no point did the supply of good quality would-be recruits slacken.

At the end of another World War the situation would be much changed. For over twenty-five years there was to be full employment, making the position of the Corps more difficult, and a German bomb would in a single act destroy both the Headquarters Barracks and canteen accommodation along with other Corps property, thereby depriving it of its traditional base.

CHAPTER ELEVEN

The Second World War and its Aftermath

On 3 September, 1939, the United Kingdom found itself at war again, less than twenty-one years after the last terrible but inconclusive conflict. For the Corps this was its third experience in forty years and for its senior management the gathering clouds had been plain for some time. Almost four years earlier the Commanding Officer had worked out the probable effects on both personnel and finance. He anticipated an inevitable reduction in strength, quite apart from the numbers who left to fight. On the day war broke out the roll stood at its record level of 5186 men. In rapid time all its reservists and many pensioners were recalled to the Colours. 1355 men rejoined, a somewhat lower figure than in 1914, for it had become an older body of men than twenty years earlier. Attempts were made to fill the posts vacated to satisfy the continuing strong demand for commissionaires. The admission rules were relaxed, men being required to deposit a flat £10 into the Savings Bank, and some considerably over the normal retirement age were allowed to join. Such over-age men could hardly join the Corps' Sick Fund and the opportunity was taken in 1939 of barring it to all new members. Its work had been accomplished and it had been made redundant due to the current National Insurance and Welfare provisions. From now onwards it would steadily diminish until it finally closed in 1975.

During his pre-war appreciation the Commanding Officer had anticipated the effects of war on the finances as a whole and appeared quite satisfied that they could stand the strain for several years, if need be, although, with the smaller income expected, some subscriptions would probably have to be raised. Economies were soon effected by the traditional wartime practice of leaving the out-divisions in the hands of their Sergeant-Majors while their officers were away and the novel one of closing for the duration half the barrack accommodation at Headquarters. There was no question this time of seeking aid from any major public fund.

In anticipation of air attack the Commanding Officer took the precaution of having built a deep underground shelter of reinforced concrete as a refuge for those men continuing to sleep at Headquarters.

In such ways did the Corps prepare itself to survive the new test.

The wisdom of the air-raid shelter became evident quite quickly, for on the night of 12 September, 1940, a heavy bomb dropped on the Barracks. Fortunately it was a delayed action device which did not explode until the following evening and the Commandant himself was able to see that all were evacuated unharmed within an hour of its landing. However, when it did explode the damage was such as to render the Barracks uninhabitable and from that time the Mess and Canteen also ceased to function.

The loss to the Corps image was considerable: the live presence of commissionaires in their Barracks was wanting and the cameraderie of the Mess and Canteen gone. On the wider scene annual parades were suspended and reunion dinners were clearly inappropriate at such a time.

The war caused other less tangible but possibly more serious damage to the organization. Less than 1200 new men were admitted during the hostilities compared with annual intakes of over 500 for the years immediately before it. The recruits themselves were unusual: while some were above age for military service, others were physically below average or men who had previously retired. This could only lead to an older, frailer body working on a diminished scale of operations in the pattern of the previous war. The roll dropped by 10%, quite apart from the 30% or so who, although still on the books, were away fighting for their country. These were administered by a staff who were themselves getting distinctly long in the tooth; the average age of the officers in 1939 was 59 and the thirty-three staff sergeants averaged 56. Little new blood could be expected here over the next few years. The Chairman of the Administrative Board, the Corps' chief decision-making body, was General Charles Lambton. He was already 82 at the beginning of the war and when he retired in 1947 he was in his ninetieth year, having served forty-one years as a member of the Governing Body and the last nineteen as its Chairman. He continued to discharge his duties with undiminished energy, and as the war was a time of increased pressure when outstanding men tended to have many extra calls on them, a successor would not have been easy to find.

The Corps' officers as a whole also found themselves burdened with extra wartime responsibilities; there was fire-watching at Headquarters and the mass of wartime regulations to cope with, including clothing coupons, new taxes, and war damage assessments. These supplemented the regular and heavy burdens concerned with the keeping of miscellaneous registers and accounts on which Edward Walter placed so much emphasis. At this time the system had changed little since his death; not until 1949, for instance, was a double-entry system instituted for the Corps' accounts. The constitution of the Commanding Officer, for one, was sorely tried by too little sleep over a long period, combined with the general strain of the London blitz and subsequent aerial attacks from V1 and V2 weapons. In 1949 he admitted that the burden of the war years had left him less energetic and he was to live for only a year after his retirement in March, 1950.[1]

One wonders how far the many irksome burdens of wartime were responsible for the Commanding Officer's remarkable paper which recommended the wholesale disposal of Corps property. According to this the only survivors should be the two properties fronting the Strand, New Zealand House and Nos

418–421, along with the Corps offices. The others he considered to be wasting assets, especially in the light of expected legislation inimical to property owners. In this hard-hitting, almost irascible paper, he first dismissed the Barracks as an anachronism, there being no moral justification for them when only 2 to 3% of the men were accommodated at subsidised rents. He then proceeded to reject any continued need for a mess or canteen which, previous to the bombing, had been patronized by only 100 regulars and a further 200–300 casuals. The property disposal would, he argued, simplify to a considerable degree 'the administrative complications from having too many sideshows tacked on to a Corps such as ours'.[2] Here was a Walter far removed from the Founder in his attitude to property. While his reference to the Barracks as being anachronistic was justifiable in the strictest sense, during the post-war years accommodation difficulties experienced by commissionaires within Central London resulted in many resignations and prevented the London Division from regaining its earlier ascendency over the out-stations. The bombing destroyed much civilian accommodation and the resultant shortage was to cause an escalation in rents which was never to be reversed. In many cases living accommodation was lost for ever, being replaced by offices and administrative units. Refurbished property owned by the Corps and made available at a fair rent for married commissionaires and their families would have doubtless been of enormous benefit. Many high-grade servicemen opted for employment with public services, where accommodation or accommodation allowance was given, when they might have joined the Corps. Many commissionaires also welcomed permanent posts with commercial organizations where accommodation was available. Suitable accommodation would possibly have given the Corps that pool of readily available manpower which it sorely needed to undertake a wider number of roles after the Second World War. Admittedly, any scheme of this nature would have been a medium-term one requiring considerable capital outlay. What Edmund Walter was also unable to anticipate, in the last years of the war, and he was not alone in this, was the great capital potential of property in Central London which became fully realizable during the '60s and '70s. Fortunately the Commanding Officer's proposals for disposal could not be implemented quickly. Although in 1947 a subcommittee, with himself as Chairman, was set up to proceed on the lines he had drawn up, the Corps' Estate Agents (Knight, Frank and Rutley) advised against the sale of the Barracks at that time.

From the end of the war retirement was always in Edmund Walter's mind. In 1946 he completed twenty years with the Corps. The problem lay in finding another Walter who was qualified (having held the combatant rank of at least Lieutenant-Colonel or its equivalent in the other two services) and was keen to assume the appointment. In October, 1947, the Commanding Officer drew the Executive Committee's attention to the fact that Captain P. Norman Walter was due to retire from the Navy in early 1948 after a distinguished career and that, although his plans were still 'in the air', he appeared most suitable as a possible successor. His record of service was thereupon recorded in the Minute Book against his possible future claims to the post.[3] In the following July Edmund Walter announced his decision to resign within eighteen months and Captain Norman Walter was invited to take up an appointment with the Corps at least six months before it was assumed he would take over as Commandant.

The war took its toll on the Corps, like so many other organizations, in both manpower and material terms. Its ranks were filled with ageing, less robust men, many of its buildings had been devastated, the income from property was much reduced and the Commandant and the bulk of its officers were approaching retirement age.

This was hardly a promising situation for the vigorous new ideas needed to meet a situation of inevitable challenge. Unquestionably the Corps' greatest contemporary need was for large numbers of good calibre younger men to offer themselves as recruits. Happily the expected rush materialized; no less than 1000 men joined in the months July to December,

1946, but owing to the large number of overage men who retired during that year, and others who left soon after joining, the increase in the total roll was a bare 100. All those recruited were placed immediately, due to the strong demand for such men, but, although wages were above the pre-war levels, so were prices. This led the Commandant to express some concern with the men's actual take-home pay after PAYE. Posts now available to commissionaires were for the most part not the best paid. In real terms they had fallen behind certain ex-servicemen's appointments since the end of the First World War and were to continue to do so. Understandably in a time of labour shortage, skilled men tended to be given priority, a fact reflected in their wages. During 1946 the number of recruits again exceeded 1000 but the large numbers of older men who continued to leave, along with others who rapidly lost their enthusiasm for the Corps, again limited total growth to 100. In his report for the year the Commanding Officer wryly observed that, owing to the ease with which men were being placed in permanent employment, the pool which had previously existed was no longer there. This was serious, since men offering themselves for temporary engagements constituted the public's main image of the Corps. Elsewhere its profile remained subdued; there were no annual parades in 1946 and 1947 due to problems with uniforms arising mainly through the restrictions imposed by the clothing coupon system.

Two years after the war, with the virtual end of demobilization, the main recruitment rush ended. The Roll had risen but it was still 6% down on its pre- war level. Henceforth it was to meet a new situation – a continually high level of civilian employment over a protracted period. In addition to full employment other external developments promised to be of considerable importance. For instance in 1947 an Interdepartmental Committee (The Wiles Committee), tasked to examine the whole range 'of problems arising on the resettlement in employment of officers and men coming out of the forces', made a number of far-reaching recommendations on the subject. Its report was based on the assumption that it was in the national interest that every man should find the employment for which he was best fitted and enter it at the level to which his qualifications and inherent capacity entitled him. To this end it recommended that negotiations should be undertaken with the Trade Unions to secure recognition of Servicemen's trade qualifications. If unskilled, men should be eligible for a vocational training scheme whereby they could take a month's training during their service and a further 3–12 months after it, at civilian skill centres where maintenance allowances could be paid during the training. Further recommendations were for Officers and Warrant Officers (the latter were the lifeblood of the Corps) to be offered a Business Training Course and for age concessions to those who wanted to enter the Civil Service.

In the event of such recommendations becoming reality, and most servicemen taking full advantage of them, it was now likely that no more than a small proportion would continue to opt for uniformed bodies like the Corps of Commissionaires. Inevitably, even if unopposed, a period would elapse between the recommendations and their implementation and, as with similar proposals in the past, some opposition or official dilatoriness could be confidently expected. Organizations like the Corps would therefore have time to reconsider their position and, if necessary, to broaden their range of employment opportunities in order to cater for a higher proportion of those men not set on skill training or undertaking managerial roles.

The short-term effects of full employment, together with a growing emphasis on men maximizing their skills and special qualities, were plain to see in respect of Corps recruiting. By the end of 1948 it had twenty men fewer than the year before and the roll was thirty-two lower still in the following year. The situation of falling numbers and the prospects of still further reductions brought problems of its own. In a time of full employment all prices tend to rise and this, together with the falling roll, meant considerable increases in subscriptions required for the clothing fund from both commissionaires and their employers. These not only produced reaction from those serving but acted as disincentives to further successful recruitment. The Corps sorely needed to re-establish its position of primacy among the ex-Servicemen's uniformed bodies, for, with the general shortage of recruits, it experienced strong competition from them, especially in London. At its request King George VI and Queen Elizabeth graciously consented to carry out an inspection in 1948 at Buckingham Palace. The inspection proved most valuable in helping to revive pride in the organization after the years of war bereft of ceremonial events. Yet the financial burden of this special parade, coming as it did upon a reduced income, meant that no annual inspection and parade could be contemplated for the next year.

The new situation experienced during the immediate post-war years was obviously of some concern to the Commanding Officer and his Executive Committee but as yet it had by no means reached serious proportions. It would have been hardly appropriate for a man contemplating retirement to have embarked on a major programme of change which his successor would have been required to implement. In fact there is no evidence from the Committee reports of the time that any such programme was ever contemplated by Edmund Walter. His inclination was to behave like most Londoners during the blitz, to continue with 'business as usual'. Even when faced with diminishing numbers, he was still most reluctant to depart from Edward Walter's original aims for the Corps. At this time the Corps had the opportunity to enlist considerable numbers of long-term employees with civilian firms who had already been carrying out commissionairing roles over some years, including the wartime ones. He rejected it. To his mind the organization 'was not merely to provide a uniform for elderly men who have for many years been well established in civil life.'[4] By any standard, though, these men could hardly have enhanced the Corps or helped the more dynamic image required of it.

While it was inappropriate and unlikely that a long-serving Commanding Officer shortly to retire, worn out by the strain of the wartime years, should contemplate any major changes, some such consideration would surely be necessary within a relatively short time. Once more it is far easier for an observer some forty years later to judge when a reassessment was needed by the Corps hierarchy than it was to see this at the time. Administrators are inevitably heavily concerned with strictly contemporary problems. At no time can an organization be in full control of its own fate; like any canoeist it tends to concentrate on the immediate problems, the hazards in the white water ahead, rather than dispassionately calculating the inexorable flow of the stream or the prospect of the wider waters beyond.

Notwithstanding, the new situation unfolding during the post-war years was sure to raise fundamental questions with respect to the Corps. Out of the reforming zeal which tends to follow major conflicts came the 1947 Wiles Committee pledged 'to achieve an easy transfer from military to civilian life', thereby solving the special and long-standing problems experienced by ex-servicemen.[5] Some ninety years earlier Edward Walter had tried to solve the same problems, but because of the strong bias against ex-servicemen in the mid-19th century his solution could go no further than the half-way house offered by the Corps of Commissionaires. The Wiles Committee's proposals were possible, partly because of his pioneering work and partly because of the changed attitude towards ex-servicemen caused by major wars and national conscription. Logically, success for these proposals would be evidence that the Corps had completed its work and henceforth could accept a steady diminution in size and function. However, while numbers remained satisfactory such a conclusion could probably never be countenanced by any contemporary Corps leadership and certainly not while the Wiles Committee's recommendations were still to be implemented fully and while considerable numbers of men clearly did not wish to take up the option of acquiring a skilled trade.

There were other basic and disturbing questions though, which had to be faced in the context of the Corps' continued existence. High standards of trust and loyalty had always been its hallmark but how far had it failed during the inter-war years to take up the changing and developing roles connected with security needs? Was there now the will, unswerving confidence in itself and the very appetite needed to compete successfully with other growing organizations and those still moving into the field? Unless it were able to extend its range of posts, the more varied and better paid positions would surely become the province of others. To offset this state of affairs could it think the unthinkable, namely that the proud title of commissionaire adopted by its Founder was no longer an unqualified asset? How exactly did the public envisage a commissionaire in the late '40s and beyond? In the public's mind a commissionaire could vary from a proud member of the Corps, through braid-laden custodians of the greater (and lesser) hotels to the attendants at local picture palaces whose main responsibility was to shine a torch on still-vacant seats. Other organizations never took on the word – the Corps would have been up in arms if they had; they were to continue to play on variations of the word security e.g. Securior, Securitas, Group 4 Security, Security Guard etc. By many counts a picture of a man performing security duties rather than 'commissionairing' seemed much more acceptable in the second half of the 20th century.

How far the Corps faced up to these questions and how it fared during the turbulent years of plenty under the last two Walter Commandants is considered in the next two chapters.

Lord Derby inspecting the Manchester Division in 1931. Lieutenant-Colonel Edmund Walter, the third Walter Commandant, is on the left of the picture.

Lord Derby takes the Annual Inspection at the Royal Hospital, Chelsea, 1933.

75th Anniversary Dinner, Shoreditch Town Hall, 31 March, 1934.

Field-Marshal Montgomery takes the Annual Inspection, 1951. Behind him is Captain P. N. Walter, the fourth Walter Commandant.

Marshall of the RAF Viscount Trenchard taking the Annual Inspection, 1952.

CHAPTER TWELVE

Full Employment: The Decade of Good Housekeeping

Captain Norman Walter became Commandant, as planned, in April, 1950, when 51 years of age. He was to hold office for ten years. From the outset he brought an air of sharp incisive administration; the Corps would be nothing less than fully shipshape during his years in office, but as with other Walters it would not change direction, continuing strictly on a course of 'steady as she goes'.

His first year in office was hardly an auspicious one. Two separate but serious challenges threatened the Corps' recruitment needs. Britain's heavy commitments to Western Defence brought about restrictions on the release of ex-Regulars during the early post-war years. In 1950 the ban was almost total and Corps strength fell during this year by 127.[1]

The second cause for concern came as a result of the Wiles Committee recommendations. These caused the creation of a further Advisory Council by the Ministry of Labour and National Service to consider servicemen's resettlement. The speed with which service skills and trades could be co-ordinated with their civilian equivalents and the integrated civilian/military career concept of the Wiles Committee realized was of great relevance to the future recruitment of the Corps. Hardly surprisingly, the process was found to be more difficult than had been anticipated and a number of advances here were to be delayed until the '60s and beyond.

This gave the Corps a partial reprieve during the 1950s and

1950 itself came to be seen as an exceptional and atypical year. Would-be commissionaires continued to come forward, if in barely adequate numbers, but, with continuing high levels of civilian employment and a greater realization of the ex-servicemen's valuable qualities, it was forced to compete in an increasingly adverse environment. More emphasis came to be placed on the quality of employment and men would leave, as never before, if they were not fully satisfied with their posts in the Corps. The needs for job-matching and for maximum job satisfaction were to become two powerful cries in the late '50s and '60s. While yet to be implemented, the ideals and objectives of the Wiles Committee were to have an immediate effect on the climate of Corps' recruiting. Serving officers concerned with advising men nearing the end of their service now took greater pains to see that, wherever possible, servicemen maximized their skills and, if they were untrained, that they considered the possibility of undertaking skill training at the growing network of Government Vocational Training Centres after their release. In a situation of full employment such advisers came to view bodies like the Corps of Commissionaires as an historic and less attractive option which preceded the 'new system of servicemen's employment-finding' they were operating.

The Corps' new Commanding Officer wasted no time in making excuses for the exceptional situation which operated in 1950 or worrying unduly about the potential effects of the Wiles Committee. He saw universal full employment as the immediate and most important problem faced by the Corps. The shortage of civilian manpower made it easy for men to obtain suitable employment, 'without joining an institution such as ours; and it is not every man who appreciates the security the Corps provides in times when it is not so easy to find a job, a state of affairs often prevalent in the past and which may well recur again'. His appreciation was undoubtedly sound, but, like so many of his contemporaries who had witnessed the depression years, he could hardly anticipate that full employment would continue for an unprecedented thirty-year period after the war.

Norman Walter was a man of action, not without the Walter quality of remoteness but someone who believed that perceived problems were there to be solved, preferably as quickly as possible. He was somewhat peremptory, not slow to acknowledge loyal support but brooking neither vacillation nor excuses from his staff nor his men. In his judgement it was essential to maintain and, if possible, to increase the numerical strength of the Corps without compromising its traditional standards. He set out to achieve this not by novel approaches but by maximizing the virtues of the present organization. Some internal reforms were long overdue but all must serve the twin objectives of helping commissionaires to greater self-respect and maximum loyalty towards their parent organization. These must make the Corps more efficient, simpler to administer, particularly with regard to its many Funds, and render it financially viable even from a somewhat reduced base.

Like Richard Walter before him, Norman was not happy about the constitution, particularly its failure to give clear directives concerning the jungle of different funds which had developed over the years and the serious legal implications which might arise in connection with them. The position of the Commanding Officer, in his opinion, also needed clarifying. The process of rewriting the constitution was long and involved, the new version not being accepted until 1955. In it the position of the Commanding Officer was now made clear and there was no question of a committee ever performing this function, although in some respects his official status vis à vis the Executive Committee had fallen still further. The President could now suspend him from duty on the recommendation of that Committee.[2] In fact under a strong Walter Commandant no material difference in relationship was apparent.

Advice was sought from the Corps' solicitors with respect to the position of the funds, but the problems here proved very stubborn and they were forced to admit that after 'extensive enquiries which we have made in conjunction with you it has been found quite impossible to ascertain with any degree of certainty the terms upon which the Savings Bank and Corps

Funds are held'.[3] However, the new constitution did go a considerable way to clarifying the position of the important Endowment Fund. Pension payments for Corps Staff were now to be authorized from it and its role was expanded in other directions to include 'such other objects in connection with or of benefit to the Corps as may be sanctioned by the Board provided such objects are not contrary to the Fundamental Principles of the Corps.'[4] One can imagine Norman Walter's frustration at the delays and difficulties he met when trying to clarify and legalize the position of the Corps Funds. From now onwards he made every attempt to wind up the Savings Bank, using monies from property disposals for this purpose.

Determined efforts were also made to cut through many of the hitherto overlapping and laborious clerical tasks which continued to be undertaken by the staff. There was a measure of success here and reductions were made accordingly in the Headquarters Staff during 1953. It was his aim to use any savings for improved recruiting measures. The need now was to attract good-quality men, whereas previously it had been to make contact with sympathetic employers. In any case the greatest deterrent to such employers lay in the Corps' inability to satisfy their manpower needs. This policy was demonstrated in 1951 when a new system of employers' subscription was instituted. It set a higher rate of contributions which increased considerably the income for the Endowment Fund and corrected the long-standing impression among commissionaires on temporary engagements that they contributed too. As from 1 April, 1954, commissionaires' entrance fees were halved and their permanent employment fees were also abolished in the hope this would give additional men some further incentive to join.

A modest publicity campaign was undertaken. Walter Commandants remained consistent in their attitude to publicity; it must never be high profile, raucous or in bad taste, although other security organizations were much less fastidious. A single advertising poster was distributed to all Service Depots and Employment Exchanges, and during 1952, a Lieutenant-Commander Oliver was engaged, for a three-month period only, to develop Corps recruiting and employment in the West Country and South Wales. Behind this initiative was the hope that the Bristol Division could be increased in strength to warrant the permanent employment of an Officer in Command there. Happily Commander Oliver was subsequently confirmed in that capacity.

As an able administrator Norman Walter was acutely aware not only of inefficiencies arising from a shortage of manpower but from the inability of using it to the best effect. The demand for labour varied from region to region. While chronic shortages might arise in some areas, other divisions might well have a surplus. He seriously considered the transfer of men from one Division to another but was frustrated here by the serious difficulties which commissionaires experienced in finding alternative accommodation. This was particularly the case in the London Division. The relative immobility of commissionaires remained a source of frustration to him virtually throughout his term, although he eventually came to recognize it as a virtually insoluble problem. He looked forward to the alternative of working up larger reserves of men available for potential employment – a long-felt need since the late 'thirties. By this means he hoped that suitable men could be made available to fill any post at short notice. Despite strenuous efforts such margins were never to be achieved, to the serious detriment of the Corps.

The Corps' property holdings were also examined with rationalization in mind. Whilst it could no longer be used for commissionaires' accommodation, it was once more proving a valuable asset. The temperament of the Commanding Officer generally favoured assets which were more flexible, though. He persuaded himself there were strong signs that punitive legislation would be passed against owners of freehold property. Major decisions were denied him in this area, however, because between the wars the most valuable part of the 'Bank Property' had been let on long leases which still had fifty or sixty years to run. This policy proved to be a most unfortunate one in view of the sharp rise in rentals after the end of the Second World War.

The Corps property had a major part to play in Norman Walter's modernization plans. In essence he was set on disposing of it whenever it was possible and profitable to do so. He saw the Corps Savings Bank as an anomaly which he aimed to run down as quickly as he could. While it was crucial in the early days to encourage thrift, he believed it had outlived its usefulness. The Post Office Savings Bank and National Savings Certificates were now available to all citizens, including commissionaires, and these paid a higher rate of interest than the Corps could afford. Repayment promised to be an expensive process; reduction of the Corps' liability towards its depositors to an average of no more than £15 a member required a capital sum of some £70,000. The only way he could raise this was through the disposal of selected property holdings. Accordingly the freehold of Nos 23 and 24 Maiden Lane (derelict properties) were sold in 1951. The Old Barracks (21/22 Maiden Lane and Burgoynes) were disposed of on a 200-year building lease and the valuable property of Walter House was to be sold for £55,000 (after considerable delays). Other properties – 4,5,6 Exchange Court, Heathcock Court and Thatched House – were let at full commercial rates and further sales were planned.

The struggle to unravel and then rationalize the Corps' finances, to dispose of certain property assets and to maximize income from the rest, to repay and then close down the Savings Fund, took an immense amount of time and energy during Norman Walter's tenure. Yet it was a most important feature of his aim to make the Corps more efficient and therefore better able to face the increased competition of the 1950s. Time was also needed for a running battle with the Inland Revenue who continued to interpret all the accounts (except the Savings and possibly the Buildings Fund) as trading funds and therefore liable to tax. This particular difficulty arose from the Founder's determination that the Corps should never be classed as a Charity, dependent as it must be on the thrift and self-help of its individual members.

In retrospect the difficulties met with when trying to dispose of Corps property, although frustrating for Norman Walter himself, were fortunate for the Corps in view of the massive escalation in property prices which was to occur over the next decade and of which the Corps then took good advantage. Even in 1952, though, after the disposals so far, the portfolio showed a significant increase over its original valuation. It was the proceeds from property in the '50s (however protracted the negotiations) which provided some of the funds for the eventual winding-up of the Savings Bank and largely helped in the modest strengthening of staff which was achieved at Bristol and Belfast.[5] Significantly Bristol was to produce most encouraging recruitment figures following the appointment of an Officer Commanding.

Norman Walter's second main objective, viz: that of helping commissionaires retain, even increase, their self-respect and identity with the Corps, could not be separated from his first. His own high standards, the loyalty and dedication of his staff, the Corps' efficiency and financial viability, the reigning monarch's agreement to stand as its Chief Life Governor, all played their part in the commissionaires' pride and regard for their organization. This was enhanced by a number of additional, if relatively unimportant, policy decisions.

Changes were initiated in respect of the uniform. During 1950 the cost of material continued to rise sharply and it was decided to change from the traditional cloth to a lighter weight variety. This was instituted in April, 1951, and it proved immediately popular as it was much more comfortable (and modern), its wearing qualities were not diminished and it retained its creases longer. In 1956 all commissionaires on joining were issued with a complete scale of new uniform so they could equal the standard of turnout achieved by the remainder and could more quickly feel themselves full members of the Corps. During the following year a new uniform cap with a white washable plastic crown was introduced. While this seemingly owed something to naval dress, it relieved the former sombre and rather dated aspect of the earlier pattern.

In respect of the traditional parades held in the grounds of Chelsea Hospital, Norman Walter felt these tended to accentuate the divisions between the VIP, the Corps Staff and the commissionaires themselves. He therefore instituted a buffet tea after the parade for commissionaires and spectators alike as an informal but appropriate way of demonstrating solidarity and Corps identity. On the debit side this caused the parades to become more expensive and from annual events they came to be held every few years, but this was something he considered preferable to more frequent parades without tea. Unfortunately the increased costs which continued to escalate with inflation led to them being held more infrequently still and finally, as the Corps diminished in strength, to cease altogether as a regular feature. Arguably they might well have been replaced by divisional dinners held annually but these never became a regular feature throughout the Corps.

While no longer annual events, during the '50s the parades continued to be splendid occasions and the Inspecting Officers were every bit as distinguished as in former years. In 1951, for instance, the Corps was honoured by Viscount Montgomery. Characteristically Monty's address drew attention to the moral force exerted by the organization, 'to help the new generation to take the right road. The young man of today needs guidance for he has to meet many new temptations arising from recent social changes and this is where the Members of the Corps . . . come in as a steadying influence.' Present was a commissionaire who had been his former drill sergeant who acknowledged with no little pride that the Field-Marshal had 'not done so badly' on the parade ground that day.

In 1952 Viscount Trenchard was Inspecting Officer; the following year elements of the Corps joined the Queen's Coronation Review of ex-servicemen in Hyde Park, while Viscount Alanbrooke stood as Inspecting Officer for the next parade four years later. The apogee of such parades came in 1959 when Her Majesty the Queen, the Corps' Chief Life Governor since 1952, reviewed it at Buckingham Palace for its centenary celebration, which nearly 1000 commissionaires attended. The final part of the Queen's message to the men on parade on this occasion went as follows: 'As Soldiers, Sailors and Airmen, you have given long and loyal service to your country; now, as civilians, you are keeping up the high standards and traditions which the public associates with your uniform. I know how much you value independence and thrift. These qualities have sustained your Corps for a hundred years. May it flourish for many years to come.'

Determined and dynamic himself, Norman Walter was most fortunate in his supporting staff, particulary his Adjutant, Brigadier Hopkins. Two Adjutants in particular rendered long and outstanding service to the Corps. One was the long-serving Major Wilkinson, Adjutant to Edward Walter, the other was unquestionably Brigadier Hopkins in his support of Norman. His enthusiasm and dedication were total and such was his personality that any employer who was considering dispensing with all or part of the Corps' services came to expect a personal visitation during which in the most polite but firm manner he was required to justify his decision. By the same token serving commissionaires thought exceedingly carefully about whether they dared leave the Corps and incur the displeasure of the Adjutant or of the more distant but equally formidable Commanding Officer.

By a succession of important administrative reforms, by sound and energetic housekeeping and by emphasis on the traditional virtues of the organization, the two men, enthusiastically supported by the Executive Committee, succeeded in maintaining the Corps' position and staunching successfully the serious bloodletting of 1950.[6] During the '50s they were even able to achieve modest gains in manpower due to a changed and increased emphasis on recruiting (arguably there was yet far to go in this respect). Radical initiatives, however, did not suit the dispositions of either man and no attempts were made during the decade to extend the range of security responsibilities or to carry the challenge to the growing number of rivals in the field.

Over the next fifteen years, under the last Walter Commanding Officer, no attempt was made to extend the Corps' security responsibilities either. Perhaps the best opportunity to do so had occurred during the '50s, for changes made in the '60s and early '70s were for the most part a reaction to pressing external events, events which were sorely to test the traditional structure and values for which the Corps had stood for so long and indeed to question its very continuance.

CHAPTER THIRTEEN

Continuing Full Employment: Against the Current

In July, 1959, Norman was succeeded by Lieutenant-Colonel Reggie Walter, grandson of Frederic and nephew to Edmund. Reggie was an ex-gunner officer like his grandfather and, like both his grandfather and uncle, he came from the side of the Walter family generally more noted for its Public Service and intellectual accomplishments than for its commercial and financial interests. Ironically, as Commandant he was to have more involvement with large-scale financial developments than any of his predecessors.

The Corps of which Reggie assumed command in 1959 was 4628 strong, only some five hundred or so lower than the all-time high reached just before the Second World War. The 1959 figure had only been reached with considerable difficulty, and by accepting a body whose average age had risen to a dangerous degree. Without a radical change in the high level of general employment there seemed to be little or no prospect of regaining that pinnacle of pre-war strength during the coming decade. In fact, if the same conditions were to prevail, there would be little hope of maintaining the levels so stubbornly held by Norman Walter. Over the next few years the Corps hierarchy came to expect that henceforth the Corps would necessarily become a smaller body. The ethics and spirit of the '60s gave little hope for growth, indeed they were to prove inimical to such a uniformed and disciplined organization. The mood of the 'Swinging '60s', that surging hypnotic beat music of the Beatles and the Rolling Stones, those hedonistic lyrics of free love, drugs and rebellion against a constraining establishment, all were in sharp contrast to the tympany of traditional brass music, the measured pace and middle-aged dignity of the uniformed Corps. Its very existence relied on the Establishment, on protecting the property of the financial houses, the professions and the major industrial concerns. It supervised activities supported in the main by the middle-aged, 'square' or property-owning class: horse racing, exhibitions, Agricultural Shows, and Services of Remembrance. Even the traditional sporting venues, large stadiums and athletics tracks which were superintended by the Corps were hardly the Meccas of the '60s trend-setters. It must have been hard for a father hanging his immaculate commissionaire's peaked cap on the hallstand to escape his children's or their own children's idolatory of ragged rebels like Che Guevara or iconoclasts like Franz Fanon. In the '60s many ex-servicemen with or without such children wanted to shed their uniforms forever on their retirement.

In the face of such pressures changes to the Corps seemed inevitable if it hoped to survive. It would be unfair to say the Corps' leaders were unaware of such contemporary trends or were unwilling to consider change at this time, but the organization's historical traditions were very strong; there seemed every reason to preserve the guiding principles of the Founder when others were losing their core values, even if the rules of the constitution might come to be interpreted more liberally. In certain quarters it was felt that if it was to go under then let it do so proudly with its values uncompromised and its main objectives of servicemen's reintegration attained. In any case Norman Walter had initiated certain areas for change and there was enough unfinished business here to take up the time and energy of both the Executive Committee and its new Commandant without looking twenty years ahead. In 1960 the Committee debated at some length changes in uniform which would bring it more into line with the spirit of the age while retaining its necessary dignity. These involved the replacement of the high-collared tunic by a jacket with a collar and tie.

Proposals were also agreed for a different and simplified pattern of crossbelt without a protruding pouch. After a referendum of all members these were implemented in 1961.[1] The modernization of the uniform crowned and completed Norman Walter's earlier attempts to help men feel themselves more at one with their Corps.

His other long-held plan, namely financial rationalization chiefly through the sales of property, proved far more difficult to achieve. Delays and setbacks here were to occur during the whole of the '60s and into the early '70s. They completely dominated Reggie Walter's time and led him to give less attention than was necessary to other important issues. The chief example here was the problem of not just a static but of a sharply falling roll. The reduction in strength came to be seen as something which was inevitable and it was met with resignation whereas there was a belief that the financial problems, difficult as they were, could be solved in time to the benefit of the organization. The decade opened promisingly enough with an increase of eighty men during 1960. Over the next three years there were steady, but as yet relatively small, decreases in strength of seventeen, forty-five and thirty-seven, followed by a minuscule rise of six in 1964. Henceforth there followed a terrible and headlong decline until 1973. The loss of men became a haemorrhage from a body which was itself ageing rapidly. If unchecked it betokened the death of the Corps in a relatively short time, whether financial viability were to be achieved or not. In 1965 the decrease was 157 men, in 1966 160, in 1967 168 and in 1972 172 men. In the early '70s the gravity of the situation was recognized by the Chairman of the Executive Committee who, in 1973, observed that if the rate of wastage continued, 'it may be that by 1990 the Corps will cease to be a viable proposition'. This was no exaggeration: during Reggie Walter's period as Commandant Corps strength declined by over a quarter, although a once-only transfer of men from the United Service Corps and the RA Uniformed Staff left its ranks only 17% below the 1960 level. The latter constituted a temporary respite only, for the transferees were themselves of high average age.

Paradoxically the minutes of the Executive Committee Meetings over the period give relatively few indications that the Corps faced such a manpower crisis or that it was prepared to take any major steps to ameliorate the situation. They were almost exclusively concerned with monetary questions: the fluid situation in respect of the various schemes for selling Corps property to developers, the investment portfolio, the possibly swingeing effects of Capital Gains Tax. All these were considered in the minutest detail. Reggie Walter and the Executive Committee, some of whom were men with considerable experience of financial and property matters, were enmeshed, even, one might say, half-bewitched, by the unfolding financial developments. In contrast, while due preliminary consideration was given to a forthcoming inspection of the Corps by Princess Marina, an inspection which came six years after the review by Her Majesty the Queen, virtually all detailed questions about its arrangements were delegated to the highly competent Adjutant, Brigadier Hopkins. Happily, the confidence was richly deserved and this turned out to be a most successful occasion. The minutes are virtually bereft of references to the recruitment problems, although regular comments were made in the Commandant's Annual Reports to the figures for recruiting and wastage.[2] But here good reasons were usually found to rationalize the adverse years. The two isolated initiatives taken over the period concerning manpower did feature in the minutes but at no great length. These were the appointment in 1966 of the retiring Adjutant, Brigadier Hopkins, to a part-time post as Employment Liaison Officer and the proposal, in 1960, of the Commandant to widen the eligibility of potential recruits. For the most part discussions were heavily dominated by the various financial plans.

Rather like the elder Cato in ancient Rome who repeated, year after year, his conviction that Carthage must be destroyed, the Commandant saw the need to debate and hopefully to implement the financial reforms set in motion by his predecessor as crucial to the efficient functioning of the Corps. For the Executive Committee, too, it became something approaching a

Holy Grail; although the financial state of the Corps was quite sound during the decade, falling numbers and inflation represented a two-fold threat which logically promised nothing less than eventual extinction. Without property sales the opportunities to raise income were limited in the main to increasing subscriptions which would tend to deplete still more the numbers of men and would-be employers. In this situation it is still debatable whether manning questions should not have been given at least equal priority with financial ones.

This fierce, almost exclusive concentration on financial matters by the Corps' leaders can only be seen as something of an imbalance. In particular there was the ironic situation of a Walter Commandant with little prior knowledge of, or present inclination for, major financial policy-making being compelled to spend the lion's share of his time on this function. It was hardly surprising that, in the circumstances, even with professional guidance, he needed to involve the Executive Committee to a degree never experienced before. If it had been his intention to delegate the unlocked funds to new and radical purposes this concentration might have been more justifiable, but this was not the case. There was a resigned, almost defeatist, air when there was consideration of the Corps' long-term chances for survival. The measures which promised to give most prospect of breaking new ground and stemming the decline were not given top priority. In reply to a question put by his predecessor on how best to employ the considerable benefits expected from financial arrangements Reggie Walter said his main ideas were to improve the officers' and staff salaries, and the officers' pension fund, to establish a free accident insurance scheme for all members, to reduce the membership subscriptions and, lastly, to provide an officer for the Liverpool Division while establishing a new division in the Plymouth/Portsmouth areas.[3] The Administrative Board thereupon adopted these as a preliminary statement of policy objectives. During 1974, after the Corps had succeeded in realizing a sum for its property beyond its wildest expectations, one still finds the Commandant addressing the Executive

Committee on the previous year's events in the following cautious and deeply pessimistic manner: 'New taxes not only undermined our financial position but incessantly increased the work load on the staff, and it was fair therefore to say that the main result of the sale of the properties would be to extend the viable life of the Corps rather than to increase its strength – though this would always be a worthwhile aim.'[4] The pervasive demands of financial questions brought another cost: the Commandant was unable to meet his members as often as he knew a leader should and he became probably the remotest of all Walter Commandants. To his great credit, though, nothing was allowed to interfere with his annual visits to the out-quarters, although the regular informal contacts between the leader and his men were much reduced.

In this situation the Adjutant had to assume greater responsibility for internal staff relations. It was, then, all the more unfortunate that some problems were experienced with this vital post. In the first six years Brigadier Hopkins continued in office. In fact he stayed two years beyond his retiring age specifically to see through the 1965 Corps Parade. While not enjoying quite the 'special' relationship which had developed between Norman Walter and himself, his personality and stature remained of immense assistance to his chief. Difficulties arose on the question of a successor. Both the Commandant and Brigadier Hopkins agreed on the choice of a Colonel Panton, but, when he was not available to take up the post, a retired Wing Commander was appointed, the first Royal Air Force Officer to join the Staff. The partnership between the Commandant and his new Adjutant was not to prove a particularly happy one. After six years his services were dispensed with and the Corps witnessed the undignified spectacle of an Adjutant attempting to sue for wrongful dismissal, but without success. As a direct result, in June, 1972, the Regulations on the Constitution and Control of the Corps, specifically in connection with the Commandant's powers to suspend any one of his officers, were subsequently amended and strengthened. Happily a successor, Colonel Pring, was

Centenary Royal Review, 1959. Marching past Her Majesty Queen Elizabeth II at Buckingham Palace.

Colour Party, Nova Scotia Division, Canadian Corps of Commissionaires, Centennial Inspection Parade Halifax, 1967.

Sergeant Kibby DSM being awarded the Corps Order of Merit First Class, after 20 years service, by Lieutenant-Colonel Reggie Walter, the last Walter Commandant.

confirmed in the post during the same month and immediately undertook his duties with distinction, establishing from the outset a close rapport with individual commissionaires, particularly those directly under his control in the London Division.

Staffing problems were the exception during the greater part of the Corps' existence, occurring with the first and last Walter Commandants only. Under the Founder some chief executives in the early years found it difficult to meet his extremely high standards and close degree of supervision. In contrast, under Reggie, engrossed as he was in financial matters, Adjutants enjoyed widespread freedom from supervision and direction. When one considers that the Corps operated under virtual siege during the '60s it did so relatively smoothly and efficiently. Because its objectives and ideals ran counter to the spirit of the age this was bound to have some effect on the morale of both its leaders and the ordinary commissionaires.

Contemporary decisions within the military establishment also posed problems. Even those initiatives for helping servicemen re-establish themselves in civilian life undertaken by the military authorities (in conjunction with the Ministry of Labour) seemed expressly designed to hamper the Corps' attempts to improve its own recruiting. Despite a period of major redundancy for senior ranks during 1957–62, the 'Golden Bowler' period when all three services were reduced by about a third, there was relatively no increase in Servicemen's unemployment, which continued to stand at a low level. Much credit here was due to the work of a Resettlement Advisory Board under the then Managing Director of Schweppes, Sir Frederic Hooper.[5] This set about improving the interviewing facilities for men and women about to leave the service. It further identified likely employment areas, provided junior management courses acceptable to potential civilian employers and, by means of resettlement committees throughout Great Britain, drummed up local support from both employers and Trade Unions. The Board built on the earlier recommendations of the Wiles Committee in attempting to exploit fully the qualities and skills possessed by senior non-commissioned officers and to find employment for them in civilian life appropriate to these skills and aspirations. This inevitably acted against the recruitment prospects of organizations such as the Corps which, while not trying to degrade a man's skills and qualities, benefited from his relative ignorance and fear of civilian life by offering a disciplined, secure and ordered organization within which he could find general employment.

It was also an age when employment of all descriptions was easy to obtain. At the end of March, 1963, from an annual outflow of 314,000 other ranks of the three services just 949 were registered as unemployed, over half of these for very short periods only. During the period it was, therefore, not surprising to find a declining trend in the number of servicemen applying to the Corps. From 627 in 1960 it had dropped to 434 in 1966, although there was a partial recovery in 1970. More serious, a great proportion of these were of advancing years, men who were not just leaving the services but ex-servicemen who had already spent many years in civilian life at other employment. Both these circumstances led to a still higher age level among Corps members than under the previous Commandant. It continued to rise. In 1964 75% of them were over 50 years of age. As it was hardly possible for men to continue serving effectively at such high age levels, there was a desperate need to improve recruiting. Two main methods of approach were possible. One could either attempt to attract men who were currently taking other employment or else relax the conditions of acceptance. In fact both lines of action were adopted but only to a limited degree. With regard to the first option the Corps could surely have taken up a much more aggressive publicity campaign, even on a limited budget. That it did not must have been partly due to the Commandant who showed the traditional Walter antagonism towards high-profile advertising and hard commercial marketing. The attempts to obtain more and different civilian posts through the personal efforts of its retired Adjutant smacked of relatively little faith in this project too. Although termed Employment Liaison Officer, he was granted expenses only for his forays and set no recruitment targets. Brigadier

Hopkins had many outstanding qualities but one does not normally start a successful marketing career in one's late sixties. His appointment was terminated by the Commandant at the end of 1972, after five years, on the grounds that 'it would save some £700 per annum and it was doubtful if his considerable efforts were producing any visible results'. Significantly no one was appointed by Reggie Walter in his place.

On the second option, the question of relaxing entrance qualifications, the Commandant put forward the most wide-sweeping proposals which would open up the Corps to all servicemen, including ex-National servicemen, men on the reserve or from the Merchant Navy, even men from 'Public Service' in general. This was thought to include all types of Government employment. He also proposed officially something which had been tacit practice for some years, that any man be admitted whom a Corps employer recommended for transfer from another part of his establishment to a work task which was normally performed by members of the Corps. This time it was the Executive Committee which drew back. General Sir Richard Gale, for one, said he thought the new proposals were too widely drawn and that it amounted virtually to opening the Corps to everybody, 'and that while it could well be possible to maintain the standards of quality, his view was that the proposal was contrary to the basic purposes for which the Corps was established and was not yet necessary'.[6] There was no denying Sir Richard's assertion that such a change would have amended the central purpose for which the Corps was established, viz: to find employment for ex-servicemen who required it. The Corps might be saved as an organization but its mission would be lost. After discussion it was agreed to admit all categories of servicemen, as proposed, but to limit 'Public Service' to mean those branches clearly comparable to the Armed Forces, e.g. the Police and Fire Service, Civil Defence Corps and Coastguards and ones with a good proportion of ex-servicemen in their ranks. Official support was not given for the admission of middle-aged men who transferred within their employers' establishment, although it was recognized that the practice would continue to take place. While this restriction succeeded in preserving the main identity of the Corps the additional categories were not large enough to check the wastage and balance the input and outflow totals. In 1970 for instance there were only twenty-two 'Others', Public Service recruits, on strength.

The plight of the parent Corps in trying to preserve both its role and organization during the adverse climate of the '60s and early '70s contrasted starkly with the situation enjoyed by its two major offspring overseas.

During 1946, in Australia, the Corps of Commissionaires (Victoria) was formed; it positively flourished during the post-war years and remains both successful and highly regarded in that State today. Edward Walter, for one, would have fully appreciated the irony that after his humiliating failure to implant a Corps at Sydney during the 1880s, after the Second World War, while the parent was itself facing considerable difficulties, an Australian Corps was to blossom, virtually unaided, in the adjoining state of Victoria.

This is closely based on the pattern of the original. As a non-profit-making concern, it is dedicated primarily to providing employment for carefully selected ex-servicemen. In some respects it goes rather further, being committed to give assistance to ex-commissionaires' widows and to veterans suffering temporary financial hardship.

Under its motto of 'Loyalty, Integrity and Service', three distinct categories of employment are offered in the traditional fields of watching and guarding, although men recruited direct from the services tend to obtain posts which are more demanding and require a wide range of specialities and skills. In the case of *Permanent* employment, men, while remaining in the Corps, become members of the businesses which accept and pay them. *Temporary* employment is offered to men who stay with an organization for a short time, but in excess of three days, while *Casual* employment involves a shorter duration still. In the latter two cases accounts for the commissionaires' services are sent by the Corps to employers.

Being restricted to a single state, the Australian Corps enjoys a very close association with that state and with the City of Melbourne in particular. It can adopt a higher profile than the parent Corps in London and the patriarchal relationship between Corps and City has enabled it to preserve traditional ceremonial functions now largely discontinued in the United Kingdom. Every year it takes part in the traditional ANZAC Day March and the subsequent Church Service in St Paul's Cathedral, Melbourne. A Meeting is held each year for all members (with a buffet meal provided) and there is a separate Annual Parade, in uniform, before the Governor, its Chief Patron. In 1983 the Governor, His Excellency Rear Admiral Sir Brian Murray, spoke about the value of such parades in positive terms,

'(They) promote standards of dress and alertness, maintain links with past service and provide the opportunity for ex-servicemen to meet and in so doing gain inspiration from kindred souls.'

Recently a Memorial Gallery was re-established to provide a good standard of rest and club facilities for commissionaires 'so that they can meet frequently and informally'. In 1984 the strength of this vibrant organization operating within the State of Victoria alone stood at 700 men.

By far the larger of the two offspring was the Canadian Corps. In Canada Government policy towards its veterans took quite another form than in the United Kingdom and one which proved to be of the greatest benefit to the Commissionaires' Corps there. Initially Canada seemed to be less than fertile ground for such an organization. Its origins date back to 1912, when a letter from E.T. Scammell, Secretary of the Naval and Military Emigration League, was sent to Frederic Walter in London raising the question of establishing a Corps in Toronto. The Adjutant provided Mr Scammell with a Corps circular but nothing more seemed to come of the request. Late in 1915, though, when there was already a prospect of large numbers of soldiers returning at the end of the war, a number of letters appeared in the Canadian press in favour of Government support for such a scheme. The Duke of Connaught, Governor-General of Canada and firm friend of the parent Corps (he was its Inspecting Officer on its Annual Parades in 1904, 1919 and 1935) also suggested the formation of a Canadian counterpart. There was another apparent lull, but in 1919 Frederic Walter received a further letter, this time from an Archibald MacMillan in Ottawa, requesting affiliation for a Canadian Corps he was about to establish in that city. Archibald MacMillan said he hoped to secure substantial Government and Public support through a Dominion Charter, as he was convinced much sympathy existed, and he went on to reveal ambitious plans to establish branches in every large city from Halifax to Vancouver along with their out-stations.[7]

There is no record of the home Corps' reply, but immediately after the war the Commandant was visited in person by a representative from the Canadian Government Department of Soldiers' Civilian Re-establishment who wanted to discuss the prospects of establishing a Canadian counterpart to the Corps. Unlike the British pattern this demonstrated firm Government support from the outset. However, on his return to Canada he presented a rather pessimistic report and the scheme was dropped temporarily by the Federal authorities. During the depression of the early post-war years serious employment problems were experienced by the older veterans and this caused a further attempt by the Canadian Government to establish a Canadian Corps, fashioned once more on the lines of the British model. A Charter was obtained and three companies were organized, No 1 in Montreal, No 2 in Toronto and No 3 in Vancouver. Once more the plant was slow to grow; all three experienced considerable problems and the idea was in danger of becoming moribund. The '30s, though, caused thousands of Canadian veterans, young and old alike, to join the long-term unemployed. Some were considered to be prematurely aged as a result of their war experiences and were give a War Veterans' Allowance, a form of accelerated old-age pension. For the majority this was not appropriate; they needed to continue working to preserve their self-respect. The

Government considered a Corps of Commissionaires the most appropriate means of providing this requirement. In 1936 the Department of Pensions and National Health decided to renew the letters patent of the original Charter if necessary, and to form a Committee to revitalize the organization. They met the Corps' Directors and jointly decided to restart the organization in Montreal prior to surveying prospects over the whole Dominion. Major-General W.B.M. King was made President of the reformed Corps and Public Funds were made available to proceed with branches throughout Canada. It was decided to control it from Dominion Headquarters, Ottawa, but to organize Divisions in every city in Canada having a population of 40,000 or over, with delegated powers through the local Veterans Assistance Committee. In June, 1937, the Governor-General, Lord Tweedsmuir, became the Corps' first Patron-in-Chief and bye-laws governing the conditions of operation were approved by the Secretary of State on 21 December, 1937.[8]

This wholehearted official support for the Canadian Corps contrasted with the situation in the Home Country where, over seventy years before, the Corps had struggled on alone relying on the self-help and thrift of its members and aid from sympathetic benefactors. Whereas in Canada the Corps alone spearheaded the resettlement arrangements, the London Corps had lost its premier place among servicemen's employment organizations before its Canadian counterpart was even formed.

Despite the immense advantages enjoyed by the Canadian Corps, expansion continued to be slow and difficult immediately before the Second World War. The breakthrough came when security work with Government Departments was offered to it, a privilege never attained by its parent organization. During the war Canadian commissionaires were in demand for the protection of war plants, airfields and public utilities but had no special concessions over performing such tasks. But in 1945 the Secretary to the Treasury Board sent a letter to all Government Departments stating that: 'The Board considers that there is work in the Public Service of Canada that could, with advantage, be performed by personnel obtained through the Canadian Corps of Commissionaires and requests that all Departments, Boards and Commissions give consideration to the employment in positions exempted from the provisions of the Civil Service Act.' From now on commissionaires were to gain extensive Government work and in 1968 the Corps had grown to approximately 7,000 men, almost double the size of the parent Corps whose strength had to be sustained in the face of widespread commercial competition and with no similar privileges for undertaking the protection of Government facilities. This role had never been considered possible for commissionaires in the United Kingdom. In 1982 the Canadian Corps exceeded 10,000 men, virtually treble the strength of its parent. The proud organization now boasted a small National Headquarters in Ottawa and eighteen autonomous Member Corps across Canada. This second offspring far surpasses the strength of its progenitor and has fulfilled, in one Commonwealth country at least, the most optimistic hopes of Edward Walter.

In one respect all three organizations shared the same problem in the early '70s, the inexorable rise in the age levels of their members, but the Canadian example showed how the parent Corps' fortunes could have differed with central Government support and the Australian body, limited as it was to one State, enjoyed a high level of sustenance there. This was not the British way, however. The outcome for their ex-servicemen in the '60s was less formalized, less supportive, but no less favourable as in the main they returned to become successful members of the greater society from which they were recruited.

With respect to the British Corps, after more than a decade of waiting, the last three years of Reggie Walter's term of office witnessed a number of important developments. After many frustrations and disappointments he was to see the long-delayed property disposal come to a startling dénouement. In the employment and manpower field the Corps was to find itself the lone survivor among ex-servicemen's uniformed employment bodies, while the demise of the Veteran's Corps and the Royal Artillery Uniformed Staff as separate organizations was to bring it an unexpected but considerable number of 'senior' recruits. In turn the conclusion of the property sale and the large capital sums which were forthcoming gave the opportunity for a redrafted constitution and the long-awaited rationalization of the different Corps Funds. As Reggie Walter was due to leave office, it was decided that responsibility for co-ordinating the financial and investment policy (under the general direction of the Corps Executive Committee) was now to devolve upon a Financial Controller. Never again would a leader be so beset with the minutiae as well as with the general conduct of Corps finances as the last Walter Commandant.

The final run-up to the property sale was by any standard an exciting and dramatic process. In 1970 the Commandant reported to the Executive Committee that the formal valuation of the Properties had just been completed and amounted to £500,000. During a later meeting of the same year Mr Roper, the Corps Auditor, informed them that in light of the Corps' current liabilities any substantial increase in income would require no less a figure than £1M cash and the grant of free office accommodation from property disposal. As yet the gap between the two figures seemed unbridgeable. By June, 1972, the Commandant was able to report that the figure he now hoped might be paid by the would-be developers could approach £1.3M, but alternative office accommodation would still need to be purchased out of this total. Six months later he announced a definite offer of £1.5M and suggested this might possibly be squeezed up. By February, 1973, offer and counter-offer had brought the figure up to £2,011,000 which the Administrative Board accepted, despite a powerful dissentient among them who questioned why the purchaser should not be talked up yet another million within a further twelve months! After a discussion which raised the spectre of a frustrated would-be developer pulling out altogether the offer was accepted unanimously. This included the facility to lease back the Corps HQ at a nominal rent for a period of five years. By this decision the Corps achieved full financial viability.

It was a triumphant vindication of Edward Walter's determination one hundred years before when, with a combination of perspicacity and tenacity, he scraped together the first modest funds to buy property, more often than not in a dilapidated condition.

One outstanding question with regard to property remained, the purchase of a new Corps Headquarters. A number of properties were seen and discarded as being unsuitable. At one point consideration was given to buying a ship and mooring it on the Thames! Subsequently, acting on the advice of Mr Trever Burfield FRICS, working on behalf of the Corps, it was decided to purchase 3 Crane Court, off Fleet Street. In February, 1974, Mr Burfield was authorized to negotiate to a maximum price of £602,000 freehold. The purchase was duly made at a time when property prices were at a temporary peak – and the cost was to rise considerably, as much work was needed to make Crane Court suitable for the Corps' purposes. The lowest tender submitted in May, 1976, for refurbishment was in the order of £148,000 and the full cost rose to £170,000. Unexpected delays were to place the heavy task of planning the accommodation requirements, and of organizing and overseeing the Corps' move to Crane Court, onto a future Commandant's shoulders. Despite the relatively high price paid for Crane Court more than sufficient capital remained from the disposal to sustain the Corps in a favourable financial position.

The adverse climate of the '60s proved equally serious for other uniformed bodies of ex-servicemen and the Corps' traditional competitors decided to cease operating by 1974 and to amalgamate with the senior body. The older and much the larger of the two was the United Services Corps formed sixty-six years before. In 1960 it was over 2000 strong (having regained the level attained just prior to the Second World War) but by the early '70s its strength fell to just over 1000. Many of its members were elderly. The Secretary told the present author in 1973 that 'quite a number of the men are around 75 years of age' and their most renowned member, CSM Bent VC MM, was 82 years of age. In 1972 only one ex-regular serviceman had applied to join, in this case an RAF Flight

Sergeant with thirty-two years service who must himself have been at least 50 years of age. The high average age of those serving, together with the acute shortage of new recruits, led to a sharply falling roll which increased the average administrative costs. Unlike the Corps of Commissionaires the Veteran's Corps owned no property and the impending expiry of its lease at Headquarters forced its decision to cease active employment. The Corps of Commissionaires gained considerable benefit from the merger, 444 Veterans enrolling in the London Division. Eighteeen months after the merger Colonel Pring, as OC the London Division, wrote to his Sergeant-Majors reminding them that, but for this addition of nearly 450 men, 'We would by now have had to be thinking seriously about some reduction in our HQ staff'.[9] This was not the whole story. The Veteran's Corps' members vastly increased the scope of the temporary employment which could be undertaken. In 1972, for instance, they had 116 permanent posts only, but carried out no less than 5787 temporary assignments. The higher echelon also gained from the merger. Its distinguished Vice-President, Major-General Sir James d'Avigdor-Goldsmid, was to become a Governor and from 1977 onwards first a stalwart member of the Executive Committee, then its Chairman in 1980.

The uniformed staff of the Royal Artillery Association, a somewhat smaller body formed shortly after the First World War, was also ageing rapidly. Initially its was expected that the greater part of 500 men in London and 150 men in the provinces would come over to the Corps. In reality this was a wild overestimate, for at the end of 1977 the number of transferees in London still serving with the Corps was down to forty-nine, twenty-six of them in permanent employment. Most of its members had proved too old for normal employment. While of relatively little significance to the Corps' own manpower problems, the merger went extremely smoothly as one would expect when the Commandant and Commandant Elect had themselves served in the Royal Artillery.

In the space of two years the Corps had achieved an enviable and permanent level of financial soundness and a much-needed breathing space in terms of manpower. Unlike the financial position, the injection of men who were for the most part 50 years of age or over could provide only a short-term benefit. New ways had to be found to check the decline in the Corps' strength if its survival was to be assured. Such thinking lay behind the third major event which occurred shortly before Reggie Walter's retirement. This was the decision to reorganize the basis of the Corps finances and, equally important, to appoint a Financial Controller, a professionally qualified man whose prime task was to relieve the Commandant from the principal financial responsibilities. Henceforth it was hoped that the Commandant could carry out his proper role which was running the Corps, in particular concentrating on the question of increasing its strength. Investments which had been held in various individual Funds were to be transferred to two Management Companies wholly owned by the General Fund and the Officers' Endowment Fund respectively. Much tidying and streamlining was still needed, but the basic framework had been established from which the Financial Controller could act. The new financial provisions also brought about amendments to the Regulations for the Constitution and control of the Corps.

Apart from legalizing the new financial arrangements, changes made to the constitution at this time formalized what had become a distinct feature of Reggie Walter's term as Commandant, namely the Executive Committee's greater involvement in detailed decision-making and its increased responsibilities.

Because of its financial suzerainty it became the practice during the years of financial embroilment for all monetary matters to be placed before it, even ones of relatively minor importance, which earlier would have fallen under the aegis of the Commandant. The Committee had long shouldered responsibility for deciding on the level of the Commandant's and other officers' salaries, and on the timing and amount of any increases. There was no question of them rubber-stamping proposals put to them in this context. For instance, in January,

1970, the Chairman of the Executive Committee reminded his members that it was the 'watchdog of Corps Finances' and that it must at no time give anybody any excuse for saying or thinking that 'The Commandant makes a good thing out of the Corps'. Contrary to what happened a hundred years before, by the early '70s the Committee made no bones about disagreeing with, and amending, certain of the Commandant's proposals if it felt it was right to do so. Queries were raised even on such matters as increases in sundry expenses or whether the sales of assets contained in the annual statements of accounts had been given the detailed attention they deserved.

It was this committee which, in June, 1974, set in motion the investigation by a firm of Management and Financial Consultants leading to the appointment of a Corps Financial Controller. The pattern of its greater and more detailed responsibilities came to take on a permanent aspect in the later '70s. In June, 1977, for instance, the chairman of the Administrative Board drew attention to the great increase in the quantity and quality of the Executive Committee's work and stressed the need to have available a sufficient number of comparatively young members to fill vacancies as they occurred. Lying behind such a statement was the consciousness of the serious and current problems facing the Corps. They were problems affecting both its structure and role. The traditional hierarchical system of a single chief executive operating without specialist support had shown itself to be inadequate in the face of the increased tempo of financial developments. After unlocking the Corps' assets, a 'professional' was required to administer them and even he would need to call on further external advice from time to time. The adverse demographic trend could not easily be changed, even when the Corps achieved financial security. A financially viable Corps might still be forced to cease functioning within a comparatively short time if this could not be reversed.

Unless the general employment situation were to change markedly any major attempts at halting the manpower losses and lowering the age structure would doubtless provide a serious challenge to the Corps' traditional role.

CHAPTER FOURTEEN

The Walter Era Ends: The Corps in the Contemporary World

Reggie Walter's retirement marked the end of an era. The Corps created and led by Walters for 116 years was visualized by its founder as the means of crossing that great divide which in the mid-Victorian era separated the serving soldier and sailor from a successful return to civilian life. It was to serve as the vehicle of his personal campaign in finding employment and opening up appropriate Government posts for such men. During the supervision of its five Walter Commandants, most of the Founder's dreams for his veterans, dreams often far ahead of his time, were to become reality not just for ex-servicemen but for the population at large. Manifestly, a private organization like the Corps, with or without assistance from *The Times*, could never have accomplished so much alone, but it could, and did, give a striking lead in certain key areas. For instance it gave a live demonstration to the City of London and to leading public figures that soldierly virtues of reliability and efficiency could themselves be of considerable practical benefit to the country's major administrative and commercial centres and to industry too. Its highly controlled workforce of erstwhile wayward men could help to justify the liberal ideals of those such as Sydney Herbert and Florence Nightingale who saw ex-soldiers as men worthy of assistance. Without such an example other military employment agencies would surely have encountered much greater opposition in developing their own wider range of 'normal' employment opportunities, agencies which themselves were to provide models for the eventual creation of civilian employment exchanges.

Notable disappointments were experienced too. In contrast with continental practice Edward Walter and the National Association which followed him were never able to progress far in persuading Governments to reserve many public posts for ex-servicemen. William Gladstone might have been a champion of the Corps Band as it played in St James's Park, but he inherited that deep suspicion of militarism and the use of force traditionally held by liberal humanists in British political and public life. Whilst it could be argued that this movement owed much to a reaction against Cromwellian excesses, it stretches forward as well to the modern proponents of disarmament from Arthur Henderson and Lord Robert Cecil, through Philip Noel Baker to Monsignor Bruce Kent. In fact the chances of significant concessions for ex-servicemen with respect to Government posts are less favourable than ever before.

Fortunately the failure of Edward Walter and subsequent champions of ex-servicemen to obtain Government posts for them was not the disaster they feared. Indeed it has played a constructive part towards that greater achievement, namely the successful return to civilian life as a whole. This has come about in the late 20th century through the serviceman's ability to find employment on the open civilian job market, employment of all descriptions and at the level appropriate to his ability and degree of ambition. In the British pattern, though, progress this far has been both piecemeal and prolonged and certain problems will always remain. Servicemen with no assurance of minor Government positions face those special and particular difficulties experienced by anyone long absent from civilian employment. Latterly the calibre of their service training and their outstanding personal qualities allied with the considerable assistance given over the transition period have enabled them to achieve a ratio of re-employment not below that of their civilian counter-parts.

Success has been due in part to developments outside the resettlement system, to structural and social changes within the armed services, the former carried out primarily for national security reasons. Some of these were not always favoured by the Victorian reformers, Edward Walter among them. A notable instance was the institution of initial short-service engagements, an unpopular development with many of the old school, but one which broke the prejudice against the long-serving, hard-bitten soldier ignorant of civilian life. Further improvements in the conditions of military life leading to greater self-motivation and enhanced personal responsibility were bound to occur when most of the male population found themselves members of the forces during two large-scale and prolonged wars.

The implementation of these, along with other civilian and military reforms, have met the majority of Edward Walter's long-term aspirations. Most notable here were the measures to provide employment for the disabled and the security afforded to the population at large, including old soldiers, by the National Health and Pension provisions. The achievement of such long-term aims has also served to break the pattern of progress enjoyed by his Corps in earlier times and to reduce its significance in the contemporary employment scene.

In strictly logical terms ultimate success for Edward Walter and his Corps of Commissionaires would suggest its winding up on the grounds that a 'half-way house' in employment, a uniformed body of men still under disciplined conrol, is no longer necessary or appropriate. In practice there remains much to do. Apart from its responsibilities to those commissionaires presently on strength, in conditions of high unemployment all employment organizations are needed, especially those which cater for servicemen who are most vulnerable, the technically unqualified. With unemployment running at a level unknown since the early '30s the Corps can and should continue to play a most important, if a more restricted, role than that conceived by its Founder over 125 years ago. As Edward Walter would have been the first to acknowledge, though, the nature of the employment provided must be expected to change. If the Corps' primary role lies in finding suitable employment for non-commissioned officers leaving the

Sergeant Henry Fitzgerald, who served in the Corps from 1934 to 1977, seen here at the offices of Morgan Crucible.

Sergeant-Major Legg with, behind, Sergeants Kirwood, Sands and Bowler; Edinburgh Tattoo, 1981.

Commissionaire Sergeant-Major John Gorman MBE, site manager for the Corps of Commissionaires, IBM Havant, 1984.

Northern Ireland, 1985. *Left to right:* Colonel Bruce Robertson, the present Commandant, Sergeant-Major Robert Cairns, Mr Douglas Hurd, then Secretary of State for Northern Ireland and Sergeant-Major Stan Burrows.

services, this should take the form of adequately paid posts with a measure of responsibility and prospects for advancement. The high-calibre men of today will be satisfied with no less and in some cases their ambitions here will never be satisfied by other, purely civilian, security organizations. The continuing and important function of temporary employment for older commissionaires in no way invalidates this primary aim.

The Corps' first non-Walter Commandant was Colonel Geoffrey Pring. He was in a unique position to understand both the distinctive nature and traditional function of the organization, having served as its Adjutant for the previous three years. On his appointment in August, 1975, he had the assistance of the new Financial Controller, Mr George Booth, while his successor as Adjutant was Colonel Alastair Thorburn. In the pattern of recent times it was to be expected that both the Commandant and his Financial Controller would work closely with the Executive Committee.

Like all Chief Executives, Geoffrey Pring, too, inherited unresolved problems, although with his experience as Adjutant he could also take up office with certain clear objectives, including a determination to exercise personal leadership and conviction about the need to review and simplify many of the long-established and laboured administrative practices. As Commandant he sought to make himself readily available in order to keep in touch with the current ideas and contemporary problems experienced by individual commissionaires. An attempt to stimulate a greater interchange of ideas among the whole staff came in 1977, with the first Conference of Outquarter Commanders. This was most successful and was to become an annual event. One practical result was the agreement to refurbish the Outquarter offices to complement the work which he was planning regarding accommodation for the Head Office staff. During his term a thorough reallocation of duties at Head Office was carried out. This aimed to relieve the Commandant from the minutiae of day-to-day administration in parallel with the reorganization which had already taken place to amend his duties in the financial field.

Two outstanding issues demanded his full and immediate attention: the impending move of Headquarters to the new premises at Crane Court and the continued and necessary rationalization of the Corps' financial position. Overshadowing everything though, was the shrinking roll of a largely elderly body of commissionaires. This itself raised a further and disturbing question, namely whether the adverse demographic trend had not gone too far to be reversed without radical decisions which must affect further both the structure and function of the Corps. The latter, however, was a prospect which would cause any Commandant, particularly a newly-appointed one, to consider most carefully before taking major remedial measures.

There were ample practical problems concerned with both the exterior and internal development of Crane Court. These were by no means straightforward, for the new premises were very much smaller than those at Exchange Court. They occupied four floors and lacked the presence created by the noble entrance hall of the original building. The result was, and had to be, a finely judged balance. There was the dual need to achieve the maximum efficiency and economy, while preserving the traditional and dignified ambience of the Corps Headquarters. In truth much had to be sacrificed despite using virtually all the appropriate wall space for the traditional Corps name boards, including those Rolls of Honour for commissionaires who, having rejoined the colours, paid the ultimate sacrifice. The presence of the Founder was certainly not wanting, though. Anyone ascending the stairs is confronted on the first landing by a large portrait which covers a whole wall. Its cool regard probably has much the same directness for a would-be commissionaire as the famous First World War recruiting poster which features Lord Kitchener. Despite the much- reduced scale of the new premises, careful and thoughtful planning did succeed in providing a Headquarters with all the facilities needed to function effectively. It was undoubtedly more cost-effective than its predecessor. Typical of the thorough and sensitive approach was the changing of necessity into a virtue with respect to the miscellany of old records, files and photographs which had accumulated in the expansive cupboards of Exchange Court for over a century. These were sorted and, after negotiation, accepted into the safekeeping of the Historical Archives of the Greater London Council.

The second outstanding issue, the struggle towards further and necessary regularization of the Corps finances, was to last much longer, continuing virtually throughout the Commandant's seven-year term and was of vital importance to the very continuation of the organization. Although the appointment of the Financial Controller, a qualified accountant, relieved him from the direct administration of Corps Funds he was still deeply involved in the area of financial policy-making at a time when grave problems continued to be encountered. So too, of course, were the Corps Executive Committee and the Financial Controller himself. The difficulty here centred on the question of taxation.

Considerable rationalization had already taken place with the Corps investments. After purchasing the new Headquarters buidling at No 3 Crane Court, the balance of the proceeds from the sale of property and of other individual funds was transferred into two Management Companies formed to hold the investments and to carry out the investment transactions for the General Fund and the Officers' Endowment Fund respectively. These portfolios of investments were now supervised by an Investment Committee which achieved a healthy capital gain after its first year of operation. However, the Corps' trading operations continued to run at a steady loss and, unless income from trading and investments could equal or exceed expenditure, it was felt the Corps' position would become most precarious. At this time the taxation status of the Corps was based on mutual trading whereby the trading results were not taxed but investment income was fully taxed. In this situation the Corps could not set off trading losses against investment income.

The Financial Controller estimated that taxation for the years 1975–7 would not be less than £50,000 and in this event the Corps might well find itself running at an overall loss despite its healthy capital position. Given its falling roll and the inadvisability of increasing subscriptions, it was decided to look for improved income by exploring ways of lightening the very heavy and seemingly inequitable tax load when compared with other employment organizations. There were two main options, obtaining charitable status for the Corps on the grounds that its work was essentially of a charitable nature, or establishing a separate charitable fund (which could receive favourable tax treatment). The chances of having the first option recognized were not aided by the traditional stance of the Founder who set his face against charitable status and had ever emphasized the fundamental principle of self-help and thrift for his commissionaires. While other military employment societies like the Regular Forces Employment Association had been granted charitable status from the outset they had never charged fees from either employees or employers. By this time the Association, though, had also come to receive further substantial assistance through grants from central funds. Legal Counsel was appointed to consider the merits of the Corps' case and an application was made to the Charity Commissioners for total charitable status. This was rejected outright. By this time, in late 1979, the Financial Controller himself had changed and Colonel L.D. Davies, late of the Royal Army Pay Corps, assumed the appointment in February, 1980, incorporating also the duties previously carried out by the Paymaster. Following rejection of the case for total charitable status, Counsel considered that some of the Corps' objects and activities might still be considered as charitable. After a detailed examination it was felt that the cost of administrative effort in differentiating between those activities which could be considered to be charitable and those which were not would outweigh any financial advantage that such part-charitable status might bring. It was, therefore, decided not to proceed with the alternative. A new approach was adopted, this time towards changing the Corps' tax status from mutual trading to that of a normal commercial venture in which trading losses could be set off against investment income. The decision here lay with the Inland Revenue and although in the first instance it rejected this proposal too a further initiative involving the Chairman of the Executive Committee, the Commandant and the Financial Controller, led to a change of heart. The Inland Revenue agreed that the Corps could be recognized as a trading venture. This offered substantial tax advantages but as the law requires a venture to intend to make a profit, the Inland Revenue qualified its agreement on the understanding it would wish to review the position in due course. In consequence of this change the Corps was restructured financially. The OEF Management Services Ltd was made dormant; the General Fund and the OEF Fund became inoperative and all financial transactions and the financial management of the Corps were channelled through the Corps of Commissionaires Management Ltd. The company in its new form began operating on 1 January, 1981, five years after Geoffrey Pring had assumed office. During the interim much energy had been expended. There had been innumerable comings and goings between himself, the Chairman of the Executive Committee, the Financial Controller, the Corps Legal Adviser and others. Ad hoc meetings of the Executive Committee had been called, papers written and innumerable verbal briefings and presentations conducted. Undoubtedly the positive conclusion of the financial question both guaranteed the Corps' continued viability and marked a high point of Geoffrey Pring's term as Commandant.

Far less successful were the attempts made to check the steady decline in membership, a trend well advanced when he assumed office. Geoffrey Pring inherited a Corps of over 3900 men; at the conclusion of his seven-year term it was just below 3000. This reduction of almost 25% compared with the same percentage loss over the previous Commandant's term of fifteen years. The reduction under Reggie Walter, however, was reduced by the winding-up of the other uniformed employment agencies and their absorption into the Corps. The statistics were not helped by the Commandant's decision during 1977 to pursue a deliberate policy of retiring those members still on strength who had become ineffective and an administrative liability, being prepared to take up neither full-time nor temporary posts. As in other respects he went for a policy of honesty and greater efficiency. His first Annual Report acknowledged the fundamental problem of Corps' recruiting which was the inability to attract younger entrants, particularly those just finishing their regular engagements. With 50% of commissionaires over 60, the wastage rate could not be reduced and the position could only be improved by attracting more men. This was far easier to state than to achieve when 'the type of uniformed work with which our name is traditionally associated commands relatively low wages and offers little opportunity for advancement.' He forecast one necessary development, namely that in the hunt for better jobs a greater proportion of non-uniformed ones might have to be accepted.

The inability to attract young ex-servicemen, and then to hold them, was demonstrated only too clearly by the figures for these years. In the last twelve months of Reggie Walter's term just 2 men joined straight from the services and during Geoffrey Pring's first year there were only 5. In 1977 the position apparently improved when 15 such men joined but most of these soon became disenchanted with the posts offered. It was back to the worst case in 1981 with two men only, although in his last full year 20 men joined. While this last increase was due to a number of approaches undertaken by the Commandant himself it also owed something to the increasingly adverse job climate in civilian life.

Geoffrey Pring clearly showed his awareness concerning the true nature of the manpower problems faced, and he recognised their seriousness, but like other Commandants before him the most he could hope for in the short term was to ameliorate rather than solve them. He instigated an external public relations campaign to keep the Corps in the public eye which resulted in a new shoulder flash and other devices to differentiate the commissionaire from other security organizations.

However, he revealed his lack of faith in the efficacy of such measures by declaring that the key to good recruiting lay rather in favourable reports from serving commissionaires to their friends and would-be recruits. This reflected his regard (and that of most serving soldiers of the time) for 'internal recruiting' but when so very few men were entering the Corps straight from the services the opportunity for them to spread the word was most limited. Simultaneously, he began to extend the pool of suitable jobs for younger and potentially more ambitious men. Some of the employment areas explored tended to involve contract work, often with large firms such as IBM. With such disappointing progress in recruiting younger men the Commandant acknowledged the reality that it was the older ones, often on temporary engagements, who were both bringing in far the larger share of revenue and flying the Corps flag most effectively. In successive reports Geoffrey Pring paid justifiable tribute to such men who, as he said, provided the best opportunities for 'our members to be seen and respected in public'. It was these men who were evident at such places as the White City, Wimbledon, the Ideal Home Exhibition or when supervising arrangements for the public display of Princess Anne's and Captain Mark Phillips' wedding presents.

Despite seemingly intractable problems in the recruiting field Geoffrey Pring's term saw important advances which owed much to the smooth coordination of an active Executive Committee and a hard-working Commandant, totally supported by an equally hard-working Financial Controller and other members of staff. Most notable here were the concessions gained over the punitive tax demands, the immaculately-planned and successful move to the new Headquarters, and the overhaul of the Corps administrative system, including the safeguarding of its historical records. With the resolution of such matters a future Commandant would arguably have more time to practise strictly command functions and to concentrate on the highest priority tasks.

The present Commandant, Colonel Bruce Robertson, assumed office in August, 1982, and straightaway faced the most intransigent and longstanding problem over the last 40 years, the Corps' diminishing strength. With the exception of the Founder no other man was to be faced so directly with the direst question of all, the survival of the organization. Time for remedial action was undoubtedly short. Unless the annual reduction in the Roll could be checked or reversed it would surely fall below 2000 by 1990 and then both the Corps' continued viability and the extent of its contribution to servicemen's re-employment would be open to serious questioning.

He has shown himself fully aware of the main challenge, consideration of which has occupied the bulk of his time so far. A twofold strategy has been adopted, an analysis for medium term action unrivalled in its thoroughness while simultaneously urgent remedial measures are being implemented.

Issues for the medium term were outlined in a paper which the Commandant presented to the Corps' Executive Committee at its meeting in October, 1984. The chief one was whether the Corps did, in fact, carry out the role originally conceived for it and, if not, what needed to be done. From his observations so far he suggested that since World War Two the organization had certainly been providing a worthwhile service for men of 50 or 60 years of age but this was not always so for younger men and women who had just left the services. Such a situation was only to be expected because over a number of years no significant proportion from this age group had offered themselves for employment with the Corps. With high unemployment the position had changed and new opportunities existed to recruit such men. Younger servicemen were coming forward and were on the waiting lists of all the Divisions but many could not be placed successfully. More still made enquiries but were deflected from joining by the majority of the posts on offer. Bruce Robertson observed that the jobs the Corps presently received were generally not of the type and quality to attract today's ex-servicemen, a conclusion reached earlier by his predecessor, Geoffrey Pring. The initial pay levels and lack of career prospects (offered by many uniformed posts for instance) were not to the liking of ambitious men. He concluded that an in-depth survey was needed for the purpose of finding better quality employment for contemporary ex-servicemen and examining the Corps' employment policy as a whole. It was anticipated that implementation of its findings would involve greater expense. Extra outgoings, for instance, would surely be necessary for professional marketing and publicity operations which would themselves rely to some degree upon a strengthening of the staff and improved office facilities both at Headquarters and within the Divisions.

The Executive Committee agreed on a thorough but rapid study of the Corps' activities and as the present staff had neither the time nor the expertise to carry out such a market survey an outside consultant was to be engaged. The man selected was a Mr Peter Loyd, an ex-serviceman himself with considerable experience of management in civilian life, whose last post was as an Executive Director with the British Institute of Management. His terms of reference were fourfold: to assess the additional types of work which could be made available to attract high quality men, to examine existing procedures between the Corps and employers and suggest improvements, to review the charges and fees raised by the Corps and to consider additional means necessary for the success of his recommendations.

By 29 January, 1985, 600 employers had been contacted, the survey validated and the Report was ready. Its main conclusions were as follows. First and most important the Corps needed to consider its objectives and functions in the contemporary world, and whether its present title reflected them accurately enough. Such a re-examination surely pointed to the need for a separation of employment functions. A precise job description was required for the duties performed by a typical commissionaire on permanent employment (and initial training given). The *orthodox commissionairing role* should be separated from one which involved *specialist security functions*. To obtain and recruit for the latter a Security Services Division under a Head of Security should be established. All recruits here would also receive initial training but they would wear a modified uniform. Both these employments should be kept separate from *Non-uniformed Specialist Staff* posts where opportunities would be provided for younger, often technically qualified men, to take up supervisory or junior management positions. For marketing purposes all administration concerned with the specialist posts should be carried out by a further separate Division.

The Survey further recommended that the Corps should take on certain of the administrative tasks currently carried out by individual employers. This was especially the case when men were to be engaged on a contractual basis. In order to perform these functions effectively there was a pressing need for standardised terms and conditions of business, including set levels of placement fees and service charges (the latter to be reviewed regularly to keep abreast with inflation). Through its greater and regular involvement here it was hoped the Corps could better ensure continuing high standards of conduct from its members and help them to be offered good starting salaries, ones that were considerably higher than those offered by many other security organizations.

In the case of temporarily employed men it recommended the average age be lowered and, if possible, a proportion of younger men should be induced to take up some of the supervisory and more senior posts in this area. Here, too, the procedures of enlistment should be amended and standardised; when members were supplied to client companies bills for their services were to be submitted direct to the firms concerned.

With regard to additional measures, there was a paramount need for the Corps to sell itself properly. (As it so justly observed inadequate publicity appeared to have been a traditional failing). A good Public Relations and Advertising Agency should be engaged and a Marketing Manager appointed to the Headquarters Staff. Individual members themselves needed to be made more aware of the need to 'sell' their Corps. Greater internal cohesion could be achieved through more frequent meetings and training sessions between Headquarters personnel and those in the out-divisions. Branches might be formed where individual members, whatever their particular employment functions, could regularly come together, develop a greater 'Esprit de Corps' and, among other activities, exchange notes about further potential job opportunities.

The Loyd recommendations served to confirm and reinforce many of the Commandant's and indeed the Executive Committee's, own reactions to the problems currently being experienced by their organization. The proposals were certainly thorough and sweeping and, like all positive findings of a good survey, some went rather further than the leadership's own thoughts.

In the main, though, they were quickly recognised as a means of helping to revitalise the Corps, and to help place it upon its traditional path, but with a fuller awareness of contemporary challenges.

Over the last twelve months a stream of new initiatives have been adopted, some as a result of remedial action, others subsequent to the Loyd Report. With the present momentum the next twelve months should see no less.

The main thrust has been both towards maintaining the levels of traditional employment posts while renewing and extending contract work, on occasion in the face of competition from other security organizations. There are now more than 80 men engaged on contractual terms and, with further growth anticipated, the Corps is now moving firmly into an important field previously left to other security firms. The special paternal interest remains, however. All men now under contract are members of an appropriately designed pension scheme; in the event of a contract being terminated after just five years they will still receive a lump sum ('nest egg'). The rise in contract employment has succeeded in another respect; it has given the Corps a welcome source of extra income and one which promises to increase further.

In the traditional field, standard employment procedures between the Corps and individual employers have now been drafted and instituted from 1st June, 1985. One now wonders why it had never been done before. These have simplified life for employers by removing previous detailed and irksome administrative tasks, but in return for its greater (standardised) clerical workload, the Corps has succeeded in raising trading income in this area too. Introduction and Annual Service fees are now charged at the rate of 10% of a man's salary in London and 5% elsewhere. In addition to the necessary administrative and supervision costs these fees include the provision and replacement of uniforms.

A Corps Advertising Agency has been officially appointed although advice had already been sought from civilian consultants over the last 18 months or so. Amended recruitment leaflets are being produced in line with the new, taut, corporate image and selective advertising is being mounted in relevant publications. Corps' representatives now attend the Services' Job Fairs conducted all over the country and frequented by considerable numbers prior to their retirement. Additionally a stand illustrating the Corps' work has featured at certain leading Business Exhibitions.

So far, despite a situation of high and sustained unemployment, these measures have not led to a definite growth situation. Indeed many have scarcely had time to take effect. They have, however, served to check dramatically the rate of numerical decline despite continued cutbacks in manpower by a proportion of civilian employers. For the first time since 1972 the Corps Roll did not fall during the twelve months to December 1984 and over 400 new members joined during that year. Happily the trend has been maintained during 1985 when 481 new members were enrolled and the strength increased by 35. All Divisions are becoming much more conscious, and indeed competitive, over their accession and retirement statistics. Like his predecessor, Bruce Robertson has been determined to reduce the numbers of ineffectives; although this has an immediately adverse effect on Corps numbers it is felt it must lead to a more efficient organization. The 'active strength deficit' has now been halved, from 147 men to 69, but happily the total Roll still holds steady.

How successful these and further measures might prove ultimately only the future can show but to the writer's mind the need first to differentiate between the different employment opportunities now being offered and then to publicise them remains fundamental. Within the space of a week he has interviewed two very different ex-servicemen. One was already a member of the Corps, the other a potentially fine recruit.

The first was one of 'Our Sergeants' in Edinburgh at the Headquarters Building, The Bank of Scotland. This gentleman, an ex-naval Petty Officer, joined the Corps quite recently after 30 years service with another civilian organization. Here he stood, upright, smart and courteous, fiercely proud of his exclusive uniform, taking immense trouble to help those who approached him. As he remarked he was glad to have a worthwhile occupation which served people and incidentally kept him fit and actively involved in society.

Two days later a different conversation occurred, this time with an Army Regimental Warrant Officer, just 40 years old, who wanted a responsible, well-paid post in the London area, where he could exercise his powers of management and his considerable experience in the security and safety fields.

Both men can surely find personal fulfilment within the Corps. But at present the second, although he represents the contemporary example of that traditional, mature recruit so highly considered by Edward Walter, would not normally approach it due both to its residual, inaccurate, but blanket image of the past and the lack of contemporary publicity to dispel it. The Corps might, as yet, have relatively few alternatives to offer him but the prevailing high level of unemployment leads him to consider their opportunities along with alternative sources of work. In the event of successful placement he would certainly become its best form of publicist both to his contemporaries and to the employer he joined, a publicist of a role no less traditional than that of his colleague in Edinburgh.

Demonstrably the work of the Corps goes on; present challenges and opportunities offer success to a leadership with the imagination and energy to seize them. Whilst no longer under the direction of Walter Commandants its contemporary leaders, ex-servicemen themselves, are no less mindful of the past, no less vigorous, no less proud of the part the organization has played and continues to play, towards the serviceman's successful return to civilian life. While not occupying its past pre-eminence, the Corps at the completion of its 125th year, when it was honoured by its Chief Life Governor with a reception at Buckingham Palace, continues with its proud mission, that of finding satisfying and appropriate work for all those men who approach it after long years of public service. Edward Walter, who always demanded the highest standard, would be pleased with, and honoured by, its achievement today.

Source Notes

Chapter One (pp 1–9)

1. *Rudyard Kipling's Verse*, Definitive Edition, Hodder & Stoughton (1912) p. 429.
2. The old soldiers were to be dressed in blue gowns with the founder's badge on their left sleeve. Kemp, Thomas, *History of Warwick and its People* (1905), 180–1.
3. Cruickshank, C. G., 'Dead-Pays in the Elizabethan Army'. *English Historical Review*, Vol 53 (June 1938), 26–7.
4. *Acts of the Privy Council of England* (New Series) XV, 99–100.
5. Grose, *Military Antiquities* Vol II (1786), 180–1.
6. The land was given in relation to its quality. A soldier with 12 pounds arrears was entitled to 20 acres in Leinster, 30 in Munster or 60 in Ulster at the end of his service. Firth, C. H., *Cromwell's Army* (1903), 18.
7. See Dean, Captain C. G. T., *The Royal Hospital, Chelsea* (1950), 35.
8. Evelyn, John, *Diary and Correspondence*, Bohn's Classical Library, Vol II (1872).
9. It was said of him, 'The meanest camp follower never sought his assistance in vain, much less a soldier'. Manners, Evelyn Walter, *Marquis of Granby* (1899), 285.
10. *Fifth Report of the Select Committee on the Army before Sebastopol* (1855), 2.
11. Anderson, Olive, 'The Growth of Christian Militarism in Mid-Victorian Britain' *EHR*, Vol 86, 1971.
12. The hymn 'Onward Christian Soldiers', the most popular of all English processional hymns, was written by Sabine Baring-Gould in 1864.

Chapter Two (pp 10–20)

1. Clifford, W. G., *The Ex-Soldier by Himself.*
2. For further details about his illustrious career see Wantage, Harriet S., *Lord Wantage VC, KCB, A Memoir* (1907).
3. *The History of the Times, The Tradition Established* 1841–6 (1939), 306.
4. Bucks Record Office Walter Papers *D/E W1/F6* Letter of John Walter II dated 10 February, 1843.
5. Greater London Record Office, Records of the Corps of Commissionaires, Box H, Letter of Edward Walter to his father dated 19 January, 1845.
6. *The History of the Times*, op cit, 11.
7. In 1878 there were strong rumours that John Walter had been offered a peerage. *The Times* became a sudden convert to Disraelian Imperialism, but he never received such an honour.

Chapter Three (pp 11–26)

1. Sir Charles Russell, House of Commons, 29 June, 1868, *Hansard*, Third Series, Vol 192, col 315.
2. Annual Report of the Army and Navy Pensioners' Employment Society for the year 1873, 8.
3. In a letter to *The Times* of 25 July, 1879, Edward Walter stated that he had for several years advocated the possibility of common action on the part of the Corps of Commissionaires and the Army and Navy Pensioners' Employment Society to make the 'road to civil appointments of the public service through the Army and Navy'.
4. 'I should not, under any circumstances, have desired to forfeit the advantages arising from self-reliance. . . . There is a satisfaction in making an effort to keep clear of pecuniary assistance, especially when so many are eager to avail themselves of it.' Edward Walter, letter to *The Times* of 2 February, 1861.

Chapter Four (pp 27–45)

1. Walter, Edward (ed), *State Rewards for State Services* (1871).
2. The Oxford Dictionary was subsequently to incorporate the new word into the English Language defining it in Walterian terms as "a pensioned Soldier employed as porter etc".

3. Edward Walter was fond of citing Sir Roger de Coverley here: "You must know I never make use of anybody to row me that has not lost either a leg or an arm. I would rather bate him a few strokes of his oar than not employ an honest man that has been wounded in the Queen's service."
4. Records of the Corps of Commissionaires, Box F, Letter of CSM to E. Walter of 27 August, 1860.
5. Ibid, Box F, 14 August, 1861; 3 September, 1861; 11 September, 1861; 26 September, 1861; 30 September, 1861 and 21 October, 1861.
6. The evidence stems from the first printed version of *The Corps' Rules and Regulations* together with a fact sheet relating to the individuals and their posts dated 1 July, 1860.
7. Florence Nightingale herself discovered that commissionaires were not perfect. She dismissed a man called Malony whom, she said, had extraordinary weaknesses 'both of body and character, who ought never to have been sent to such a situation as mine at all'. Yet she appealed to the Founder not to dismiss him from the Corps for the sake of his family. Records of the Corps of Commissionaires, Box F, letter of Florence Nightingale to EW 13/1/69.
8. Records of the Corps of Commissionaires, Box F, letters of 22, 23 October, 1861. The Corps Rules of February, 1859, acknowledged the practice by including a sentence stating that 'every instance of wilful overcharge will be severely dealt with, and on a second case being reported the offender will be liable to be dismissed'.
9. Rules and Regulations of the Corps of Commissionaires, February, 1859, paragraph 12.
10. One such Article referred to appeared in the 1 September, 1866, edition of the *Comet*.
11. Records of the Corps of Commissionaires, Box F, Letter of Gordon to Edward Walter dated 5/7/71.
12. Rules and Regulations of the Corps of Commissionaires, March, 1864, p 14.
13. Rules and Regulations of the Corps of Commissionaires, February, 1859, para 12.
14. Edward Walter's emphasis on financial detail remained even when the financial situation had materially improved. In 1880 he wrote on the question of a particular Commissionaire's entrance

money, enclosing the Corps rules to illustrate his points. At the end of the letter he asked his correspondent to remit the price of the pamphlet containing the rules to the Adjutant, if he intended to keep them!

15. Records of the Corps of Commissionaires, Box D, Monthly Circulars.
16. The formula whereby the engagement money was shared between the musicians was tortuous and complicated but, in the way of the Founder, it covered every possible eventuality.

Chapter Five (pp 46–59)

1. Records of the Corps of Commissionaires, Box FF, Report of the Commanding Officer for 1864.
2. As a result of *The Times*' need to be assured how the income from the Crimean Fund was spent, the Manager was elected a Governor of the Corps 'for the time being'. It was to prove a permanent arrangement.
3. This was given full treatment in *The Times* of Saturday 5 September, 1868.
4. Sir John Burgoyne had been appointed a Trustee of the Corps in the same year.
5. This offer was welcomed by the Government and the Bow Street Magistrates accepted the help of several men by day and night. However, to Edward Walter's evident disappointment, they did not use the London Division's '200 picked and able-bodied soldiers' as a unified force. Records of the Corps of Commissionaires, Box D, Report of the Commanding Officer for 1868.
6. Records of the Corps of Commissionaires, Box AD, Minutes of Executive Committee Meeting for 1875.
7. Records of the Corps of Commissionaires, Box ZZ, letter of EW to Messrs Darley and Cumberland of April, 1889.
8. Memorandum of 13 February, 1876.
9. The sources and amounts were as follows:—

1. Duchess of Westminster	£6,721	0.	0.
2. Officers or ex-officers from HM Army or Navy	3,132	15.	2.
3. From regimental subs of £25	2,637	14.	7.
4. From the public in general	1,941	10.	0.
	£14,432	19.	9.

Records of the Corps of Commissionaires, Box Z, Circular letter of EW dated 31 March, 1876.

10. The figure was arrived at by comparing the weekly wage of 21/– for the average commissionaire at the time with the present average wage of a manual worker at £149.
11. Records of the Corps of Commissionaires, Box D, Report of the Commanding Officer for 1872.
12. Records of the Corps of Commissionaires, Box D, Report of the Commanding Officer for 1875, 6.
13. Its full title was '*State Rewards for State Services, being an Epitome of Opinion of civilians and naval and military officers on the employment in subordinate situations of the Civil Service of Soldiers and Sailors,*' 1871.
14. *State Rewards for State Services*, pp 16, 17.
15. The expenses incurred during the Meeting held at the RUSI to discuss civil employment for ex-soldiers were in fact defrayed by the Corps of Commissionaires.

Chapter Six (pp 60–72)

1. Records of the Corps of Commissionaires, Box AD, Financial Statement for 1 January to 31 December, 1885.
2. Records of the Corps of Commissionaires, Box AD, Letter of Major-General Cameron, Commanding Northern District, of 5.8.1882.
3. Letter of Major-General Cameron of 5.8.1882.
4. Records of the Corps of Commissionaires, Box FF, Commanding Officer's Address at Annual Parade.
5. Records of the Corps of Commissionaires, Box ZZ, Letter of Edward Walter, of 10 November, 1880, to Colonel Dunbar.
6. Records of the Corps of Commissionaires, Box AD, Report of the Commanding Officer to the Executive Committee for 1880.
7. PRO, Midleton Papers 30/67/3, letter of C in C to St J. Broderick dated 12 February, 1887.
8. Headquarters Corps of Commissionaires, Roll Books.
9. Records of the Corps of Commissionaires, Box ZZ, Letter of Edward Walter, of 3 October, 1884, to Colonel Brind.

Chapter Seven (pp 73–83)

1. RA ADD MSS E1/0054, letter of General Sir D. M. Stewart to HRH of 22 August, 1884.
2. RA ADD MSS E1/10906, letter of HRH to General D. M. Stewart of 22 September, 1884.
3. The History of the Association. Unpublished Monograph held at the Head Office of the Regular Forces Employment Association (formerly National Association for the Employment for Reserve and Discharged Soldiers), 1.
4. Records of the Corps of Commissionaires, Box AC, Circular of E.F.C., Simla, 22 October, 1884.
5. RA ADD MSS E1/10979, letter of Lord Hartington to HRH of 10 December, 1884.
6. Greater London Record Office, Records of the Corps of Commissionaires, Box AC.
7. Op cit, Box ZZ, letter of Edward Walter to Major Boyton of 28 March, 1889.
8. Headquarters RFEA, Report of the First Annual Meeting of the NAERDS (1886).
9. In 1896 the Pensioners' Employment Society found no less than 2353 vacancies, still well over half the combined figures of the Regimental District Registers and the National Association offices which totalled 3815.
10. Headquarters Corps of Commissionaires, *Roll Books*.
11. Headquarters Corps of Commissionaires, CG 123, Letter of Sir Charles Phipps to Lt General Sir G. Wetherall (4.8.1859).
12. Records of the Corps of Commissionaires, Box D, Annual Circulars.

Chapter Eight (pp 84–95)

1. Records of the Corps of Commissionaires, Box SS, *Sydney Echo* of 19 December, 1886.
2. Ibid, Box SS, *Sydney Morning Herald* of 27 April, 1889.
3. John Springhall, *Sure and Steadfast, A History of the Boys' Brigade 1883–1983* (1983) pp 76–93.
4. Criticism of shortcomings in the system helped him to sell his work. 'Shillin'a Day' was sent to and published by five different magazines in December, 1889.

5. *The Leisure Hour*, March 1885.
6. Records of the Corps of Commissionaires, Box YY, Cassell's *Saturday Journal* of 1877.
7. Records of the Corps of Commissionaires, Box G, Letter to HM Office of Works 16.6.1875, Letter of EW in *The Times* of June 27, 1875.
8. Records of the Corps of Commissionaires, Box 22, Letter of EW to Inland Revenue.

Chapter Nine (pp 96–109)

1. Headquarters RFEA, Seventeenth Annual Report of the NAERDS (1902), 7.
2. They both tended to revert to their original aims within a relatively short period and came to delegate employment-finding to the National Association. Yet in 1905 the Help Society placed 4704 men, although its employment responsibility was to diminish steadily from this time onwards.
3. It also opened four branches in South Africa, at Johannesburg, Cape Town, Bloemfontein and Pietermaritzburg, to help soldiers who wanted to settle there due to the difficulties experienced by men in finding jobs at home.
4. It was of course a fundamental principle that the Officers' Endowment Fund could not be subsidized by the Corps as a whole. This made any question of mortgaging the assets for rapid growth a virtual non-starter.
5. Records of the Corps of Commissionaires, Box D, Report of the Commanding Officer for 1900.
6. Archives of *The Times*, photocopy library. Article on the death of Sir Edward Walter 26/2/04.
7. Records of the Corps of Commissionaires, Boxes AE, FF.
8. Whitehall Library, Pedley A. C., *Notes on the Days that are Past* (unpublished autobiography), 116.
9. Headquarters RFEA, Twenty-Fourth Annual Report of the NAERDS for 1908–9, 12.
10. Clifford W. G., *The Ex-Soldier By Himself* (1916), 74, 75.
11. Records of the Corps of Commissionaires, Box D, Report of the Commanding Officer for 1908.

Chapter Ten (pp 110–126)

1. Records of the Corps of Commissionaires, Box AE, Letter of 5 July, 1915, from National Relief Fund to Major-General Sir L. V. Swaine.
2. Records of the Corps of Commissionaires, ibid, Letter of R. Ellis Cunliffe, Solicitor Board of Trade, to Messrs Soames, Edwards and Jones of 1 January, 1919.
3. Interdepartmental Committee under Sir Malcolm Ramsay on 'Employment of Disabled Men in Government Departments' (1919) (Disabled men incapable of normal work in Lord Roberts Workshops and British Legion).
4. Cabinet Committee under C. Addison MP, on Cooperation of Government Departments in the employment of ex-Service men in the Building Industry (1921).
5. War Office Committee under Colonel E. J. L. Pike (AGU) on Re-settlement of Ex-Regular Soldiers in Civil Life.
6. Records of the Corps of Commissionaires, Box C, Commanding Officer's Report for 1922.
7. Records of the Corps of Commissionaires, Box G, 'Notice to Gentlemen employing Commissionaire Cabs'.
8. *The Times* of 1 July, 1932.

Chapter Eleven (pp 127–136)

1. Aware of his waning energy Edmund Walter made a special request for the retiring Paymaster, Colonel Macpherson, to be appointed as his personal assistant in the years immediately after the war.
2. Records of the Corps of Commissionaires, Book AE, Memorandum to Governors on the Corps Property by Lieutenant-Colonel Edmund Walter on 20 July, 1944.
3. Records of the Corps of Commissionaires, Book AE, Notes of Executive Committee Meeting 30 October, 1947.
4. Records of the Corps of Commissionaires, Box C, Commanding Officer's Report for 1947.
5. Report by the Interdepartmental Committee on Resettlement in Employment of Members of the Forces (Wiles Report) p 1.

Chapter Twelve (pp 137–146)

1. Records of the Corps of Commissionaires, Box C, Commanding Officer's Report for the Year 1950, para 2.
2. Headquarters Corps of Commissionaires, Minute Book, Constitution of the Corps approved by Administrative Board on 23 March, 1955.
3. Ibid, Letter of Soames, Edwards and Jones, 18 June, 1952.
4. From 1954 onwards the Accounts for the Officers' Endowment Fund and the Administrative Fund were presented in a consolidated form and audited by the Corps' Chartered Accountants.
5. Records of the Corps of Commissionaires, Box C, Report of the Chairman of the Executive Committee for 1954.
6. Norman Walter even found time to refurbish the Corps' image in Brookwood Cemetery. In 1952 the headstones were reset and the obelisk to the Founder repaired.

Chapter Thirteen (pp 147–165)

1. HQ Corps of Commissionaires, Minute Book, Annual Meeting of Administrative Board for 5/7/1961.
2. Records of the Corps of Commissionaires, Box C, Annual Circulars.
3. Minute Book 1963–9 Minutes of the 1963 Annual Meeting of the Administrative Board.
4. HQ Corps of Commissionaires, Minute Book 1970–77 Minutes of the 1974 Annual Meeting of the Administrative Board.
5. Resettlement Advisory Board Progress Report 1957–9 (The Hooper Committee). This was presented to Parliament in July, 1959.
6. Minute Book 1970–77 Minutes of 277 Meeting of the Executive Committee.
7. Records of the Corps of Commissionaires, Box C, Letter of 27 January, 1919, to Major Frederic Walter.
8. HQ Corps of Commissionaires unpublished monograph, *A Brief History of the Corps in Canada*.
9. HQ Corps of Commissionaires, Letter of Colonel Pring of 20 January, 1975.

Index